PLANT SCIENCE RESEARCH AND PRACTICES

JASMONIC ACID

BIOSYNTHESIS, FUNCTIONS AND ROLE IN PLANT DEVELOPMENT

PLANT SCIENCE RESEARCH AND PRACTICES

Additional books in this series can be found on Nova's website under the Series tab.

Additional e-books in this series can be found on Nova's website under the e-book tab.

PLANT SCIENCE RESEARCH AND PRACTICES

JASMONIC ACID

BIOSYNTHESIS, FUNCTIONS AND ROLE IN PLANT DEVELOPMENT

LUCILLE MORRISON
EDITOR

nova publishers
New York

Library of Congress Cataloging-in-Publication Data

Jasmonic acid : biosynthesis, functions and role in plant development / editor: Lucille Morrison.
pages cm. -- (Plant science research and practices)
Includes index.
ISBN 978-1-63463-854-8 (hardcover)
1. Jasmonic acid. 2. Plant regulators. 3. Plants--Development. I. Morrison, Lucille, editor. II. Series: Plant science research and practices.
QK745.J38 2015
571.7'2--dc23

2014048430

Published by Nova Science Publishers, Inc. † New York

CONTENTS

PREFACE

Jasmonic acid (JA) is an essential phytohormone for plant development and stress response. It not only acts as the regulatory molecule in root elongation, germination, pollen production, fruit ripening and plant senescence, but also acts as the signaling compound activating plant defenses against pathogens, herbivory, wounding and abiotic stress. This book discusses the biosynthesis, functions and the role jasmonic acid has in plant development.

Chapter 1 – Compatible plant-pathogen interaction results in various hormonal level changes, triggering a vast array of genes to defend from the invading pathogen. Along with the phytohormones salicylic acid (SA) and ethylene (ET), jasmonic acid (JA) regulate both basal and resistance (*R*) gene-mediated defense responses. There are many evidences that JA and ET signaling work synergistically whereas SA and JA/ET signaling is by an antagonistic cross-talk. Contrarily, it has been shown that both salicylic acid (SA) derivative methyl salicylate (MeSA) and methyl jasmonate (MeJA) are essential for systemic resistance against Tobacco mosaic virus (TMV), possibly acting as the initiating signals for systemic resistance. Silencing of SA or JA biosynthetic and signaling genes in *Nicotiana benthamiana* plants increased susceptibility to TMV. Locally increased concentrations of JA were observed in response to pathogen infection or tissue damage and its exogenous application induced the expression of defense related genes. Induced resistance response against necrotrophic fungal pathogen and arthropod herbivory was also achieved by JA treatment in tomato seeds. Furthermore, foliar application of JA followed by SA triggered the strongest systemic resistance against TMV. In this scenario, JA signaling plays a prominent role in plant defense responses and some important research insights in this field will be discussed in the chapter.

Chapter 2 – Jasmonic acid (JA) is an essential phytohormone for plant development and stress response. It not only acts as the regulatory molecule in root elongation, germination, pollen production, fruit ripening and plant senescence, but also acts as the signaling compound activating plant defenses against pathogens, herbivory, wounding and abiotic stress. Determination of JA has been a crucial issue in plant research. Like all the other plant hormones, JA occurs at a trace level in a very complex matrix. So it is a challenging task to achieve the complete isolation and sensitive detection of JA. In the past few years, lots of researchers focus on this particular topic to develop various methods for JA separation and detection. The ultrasensitive and in-situ detection of JA has been achieved thanks to the latest advances in chromatography and mass spectrometry technologies. Nowadays even accurate analysis of JA stereoisomers has been achieved, offering a powerful tool to understand JA's stereospecific bioactivities. This paper mainly presents the recent development in separation and determination of phytohormone JA. According to diversified.

Chapter 3 – Jasmonic acid (JA) is a signal molecule of oxylipin nature, a classic participant of plant reactions to biotic stress coordinating their defense responses against necrotrophic pathogens and pests. The role of JA in regulation of induced systemic resistance (ISR), caused by beneficial microorganisms, the so-called plant growth-promoting rhizobacteria (PGPR), is well known. Development of ISR is manifested in such defense responses as deposition of callose, lignin, and phenolics at infection sites, increased activities of chitinase, peroxidase, polyphenol oxidase, and phenylalanine ammonia lyase (PAL), enhanced phytoalexin production, and induced expression of stress-related genes. This is likely to occur due to ability of PGPR to secrete a broad spectrum of metabolites, one of them being JA. Unfortunately, the mechanism of ISR regulation and the role of JA in it are not sufficiently studied. Nevertheless, the importance of reactive oxygen species (ROS) in ISR induction under the influence of elicitors of bacterial nature is suggested.

The present report considers the aspects of the action of JA and entophytic bacteria *Bacillus subtilis* on the functioning of plant defense responses against fungus pathogens associated with activated transcription of PR-protein genes, intensity of accumulation of hydrogen peroxide, activation of peroxidase and lignin deposition in cell walls. It is shown that combined effects of endophytic strain of *B. subtilis* 26D and JA on defense response of wheat plants against hemibiotrophic and necrotrphic infection vary from stimulation to complete inhibition depending on concentration of signal molecule and succession of its application.

Chapter 4 – Plant growth regulators are not only involved in regulating developmental processes in plants but also play important roles for the plant's responses to biotic and abiotic stresses. Jasmonic acid (JA) is a naturally occurring plant growth regulator biosynthesized from linolenic acid by the octadecanoid pathway. The major function of JA is to regulate plant growth responses to abiotic and biotic stresses as well as plant growth and development. Systemic acquired resistance (SAR) is a potent innate immunity system in plants that is effective against a broad range of pathogens. Accumulation of Jasmonic acid (JA) and Salicylic acid (SA) in non-infected part renders the whole plant more resistant to subsequent pathogen infection. Mobile signals for systemic acquired resistance are translocated via vasculature and may include small peptides such as methyl salicylate, jasmonate, azelaic acid and other lipid-derived molecules. The JA signal pathway involves several signal transduction events: the perception of the infection or stress stimulus and transduction of the signal locally and systemically; the perception of this signal and induction of JA biosynthesis; the perception of JA and induction of responses; and finally, integration of JA signaling with outputs from the SA, ethylene, and other signaling pathways. Although the JA, SA and ethylene signaling pathways are clearly defined by the signaling molecules they synthesize, they interact cooperatively and antagonistically in a variety of responses. This chapter focuses on biosynthesis of Jasmonic acid and its role in whole plant resistance or systemic induced resistance.

In: Jasmonic Acid
Editor: Lucille Morrison
ISBN: 978-1-63463-854-8

Chapter 1

JASMONIC ACID: MEDIATED PLANT DEFENSE

Maria Eugênia Lisei de Sá*[1,2]*,
Magda Aparecida Beneventi*[1]*,
Fabrício Barbosa Monteiro Arraes*[2]*,
***Regina Santos de Amorim*[2]**
***and Maria Fatima Grossi–de–Sa*[2,3]**
[1]Agricultural Research Company of Minas Gerais State, Uberaba, Brazil
[2]Embrapa Genetic Resources and Biotechnology, Brasilia, Brazil
[3]Catholic University of Brasilia, Brasilia, DF, Brazil

ABSTRACT

Compatible plant-pathogen interaction results in various hormonal level changes, triggering a vast array of genes to defend from the invading pathogen. Along with the phytohormones salicylic acid (SA) and ethylene (ET), jasmonic acid (JA) regulate both basal and resistance (*R*) gene-mediated defense responses. There are many evidences that JA and ET signaling work synergistically whereas SA and JA/ET signaling is by an antagonistic cross-talk. Contrarily, it has been shown that both salicylic acid (SA) derivative methyl salicylate (MeSA) and methyl jasmonate (MeJA) are essential for systemic resistance against Tobacco mosaic virus (TMV), possibly acting as the initiating signals for systemic resistance. Silencing of SA or JA biosynthetic and signaling genes in

Nicotiana benthamiana plants increased susceptibility to TMV. Locally increased concentrations of JA were observed in response to pathogen infection or tissue damage and its exogenous application induced the expression of defense related genes. Induced resistance response against necrotrophic fungal pathogen and arthropod herbivory was also achieved by JA treatment in tomato seeds. Furthermore, foliar application of JA followed by SA triggered the strongest systemic resistance against TMV. In this scenario, JA signaling plays a prominent role in plant defense responses and some important research insights in this field will be discussed in the chapter.

INTRODUCTION

Plants are constantly being invaded by an array of pathogens like, virus, fungus, bacteria, nematodes and insects that decrease plant growth and reproduction. These pathogens are broadly divided into biotrophics (those that maintain a living host to continue their life cycle), necrothophs (those that kill the host and feed on the vital compounds) and hemibiotrophs (those which require a living host initially, but kill at later stage of infection) (Hammond-Kosack and Jones, 2000).

Despite the existence of a variety of pathogens, only a few succeed in causing disease. This is possible due the development of a sophisticated innate system by the plants, which can be divided into two categories. The first one recognizes and responds to molecules common to many classes of microbes, including non-pathogens (Microbial-Associated Molecular Patterns-Triggered Immunity - MTI), and the second responds to pathogen virulence factors (Effector-Triggered Immunity - ETI), either directly or through their effects on host targets (Jones and Dangl, 2006) (Figure 1). MTI confers basal resistance, while ETI confers durable resistance, often resulting in rapid programmed cell death called the hypersensitive response (HR) (Hammond-Kosack and Jones, 1996). The activation of HR responses can be followed by the establishment of the systemic acquired resistance (SAR) throughout the plant, which is long lasting and effective against a broad spectrum of pathogens (McDowell and Dangl, 2000). SAR is characterized by the enhanced expression of defense genes such as pathogenesis-related (PR) genes like PR-1, PR-5, and BGL-2 in the systemic tissues (Gupta et al., 2000). Plants are also able to express induced defense at sites distantly located from the initial infection, resulting in enhanced systemic resistance to subsequent pathogen attack (Zhu et al., 2014), which is referred as induced systemic resistance (ISR) (Durrant and Dong,

2004). In plants, ISR is correlated with accumulation phytoalexins and increased activity of PR proteins such as chitinases, β-1,3-glucanases and peroxidases (Peer et al., 1991; Dalisay and Kuc, 1995).

The complexity of the resistance mechanism suggests that highly integrated signaling pathways are involved in the stress perception and transduction, including changes in the level of various phytohormones like jasmonic acid (JA), salicylic acid (SA), ethylene (ET) and abscisic acid (ABA) (van Wees et al., 2000; Balbi and Devoto, 2008). SAR responses are associated with the accumulation of salicylic acid (SA) (Durrant and Dong, 2004), while ISR is dependent on JA and ET (Pieterse et al., 1998).

This review summarizes some reports on mediated-JA defense responses and its crosstalk with different signaling pathways employed by the plants to tackle pathogenesis.

PHATOGEN RESISTENCE MEDIATED BY JASMONIC ACID

1. JA/SA Crosstalk-Mediated Defense: Synergist and Antagonist Responses

Jasmonic acid (JA) and its volatile methyl ester (MJ), collectively termed jasmonates, are lipid-derived cyclopentanone compounds that occur ubiquitously and exclusively in the plant kingdom (Pirbalouti et al., 2014). Among the plant hormones JA and related derivatives are known to mediate biotic and abiotic stress responses, in particular herbivory and wounding, and several developmental processes, such as seed germination, root growth, flowering, fruit ripening, and senescense (Cheong and Choi, 2003; Delker et al., 2006). Although JA pathway is critical for signaling pathway mediating defense in plants, it requires a fine-tune crosstalk with other signaling pathways, such as those for ethylene (ET), salicylic acid (SA) and abscisic acid (ABA) (Lorenzo and Solano, 2005) that provide the plant with a such a powerful regulatory potential (Koornneef and Pieterse, 2008).

Induced resistance is regulated by a network of interconnecting signal transduction pathways that triggers hormonal changes to modulate disease and resistance dependent on pathogen lifestyles and the host genetic constitution (Robert-Seilaniantz et al., 2011). JA/SA is one of the best examples of crosstalk-mediated defense in response to pathogen infection or herbivore damage, resulting in the activation of distinct sets of defense-related genes and SAR (Spoel et al., 2003; 2007).

Signaling interactions can be either mutually antagonistic or synergistic, resulting in negative or positive functional outcomes (Koornneef and Pieterse, 2008).

It has been suggested that effective defense against biotrophic pathogens are regulated by SA-dependent pathway (Glazebrook, 2005). In contrast, pathogens with a necrotrophic lifestyle and herbivorous are resisted more through JA/ET-mediated defenses (Thomma et al., 2001; Kessler and Baldwin, 2001; Glazebrook, 2005). During HR, JA is produced and triggers the JA-dependent lesion containment, antagonizing the cell death processes elicited by SA, ethylene (ET) and ROS, through SA suppression, thus preventing excessive damage to host tissues (Overmyer et al., 2003) (Figure 1).

An example of a negative interaction between JA and SA is the silencing of phenylalanine ammonia lyase (PAL) in tobacco, an enzyme synthesizing precursors to SA, which reduces SAR to *Tobacco mosaic virus* but enhances herbivory-induced systemic resistance against *Heliotis virescens*. Contrarily, overexpression of PAL revealed an inverse relationship between SA and JA levels, i.e., enhanced SAR but reduced wound response after insect herbivory attack (Felton et al., 1999).

It was also demonstrated that during plant-*Botrytis cinerea* (a necrotrophic fungus) interaction, *A. thaliana* and tobacco plants defective in SA signaling were more resistant to botrydial (a phytotoxic sesquiterpene produced by *B. cinerea*) than wild type plants, as opposed by *A. thaliana* defective in JA signaling, which were more sensitive (Rossi et al., 2011).

Activation of gene transcriptions by hormone signaling, such as JA and SA, is dependent of ubiquitination machinery to trigger the degradation of transcriptional repressors and quickly to activate genes in plants upon pathogen attack (Robert-Seilaniantz et al., 2011). In elicited cells, JA conjugated with the aminoacid isoleucine (JA-Ile) is recognized by an F-box protein (COI1 – coronatine insensitive1) that functions as the substrate-recruiting module of the Skp1-Cul1-F-box protein (SCF) ubiquitin E3 ligase complex (Sheard et al., 2010) (Figure 2). It has been demonstrated that during up-regulated JA-Ile synthesis, JA stimulates the specific binding to JASMONATE ZIM (JAZ) family of repressor proteins leading to the degradation (ubiquination) of JAZ by the 26S proteasome (Chini et al., 2007). Destruction of JAZs activates various transcription factors, such as Myc and Myb, triggering defense-related pathways and altering the plant growth and development (Pauwels and Goossens, 2011).

Moreover, it was found another important controller in JA-mediated defense responses, the jasmonate-associated VQ motif protein (JAV1) (Hu et al., 2013). Similar to JAZ, JAV1 is degraded via the 26S proteasome and its disruption releases target transcription factors from repression for downstream JA responses.

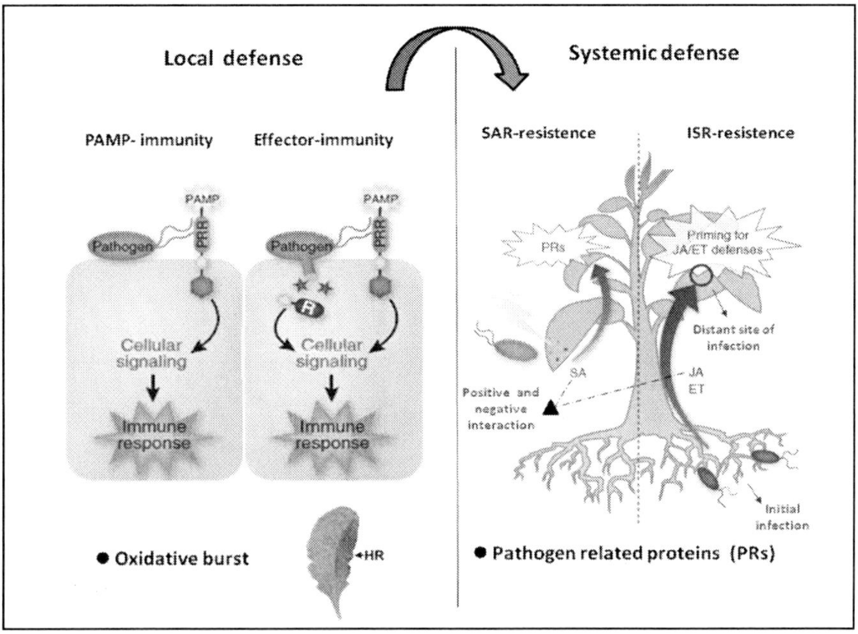

Adapted from: Pieterse et al., 2009.

Figure 1. Plant inducible defense mechanism. Upon pathogen attack plants utilize a wide variety of defense mechanisms to trigger rapid and effective responses. The host activates a signaling pathway that leads to PAMP-triggered immunity (PTI) after the recognition of pathogen-associated molecular patterns (PAMPs) by pattern-recognition receptors (PRRs). Furthermore, attacker-specific effectors can be recognized by plant resistance (R) proteins resulting in effector triggered immunity (ETI). Once a defense response cascade is initiated it often culminate in a hypersensitive response (HR) consisting of localized cell necrosis at the infection site. The systemic defense is then activated in local and distant site of infection to protect tissues against subsequent pathogen attack by systemic acquired resistance (SAR) or induced systemic resistance (ISR). Salicylic acid (SA) regulates several genes that encode pathogenesis-related proteins (PRs) during SAR while jasmonic acid (JA)- and ethylene (ET)-dependent signaling pathways regulate ISR. However, JA and SA signaling interaction can be either mutually antagonistic or synergistic resulting in effective defense during the complex interaction of plants with their attackers.

However, JAV1 has no apparent role in plant growth and development and its control seems to be related only to defense responses. These authors showed that JAV1-RNAi *Arabidopsis thaliana* plants exhibited enhanced resistance against necrotrophic pathogens and herbivorous insects without influencing plant growth and development. Furthermore, defense-related genes in JAV1-RNAi plants were induced more strongly by JA and wounding treatments while most SA-responsive genes remained unchanged. This indicates a specificity of JAV1 role in JA-mediated defense.

Contrasting with the proposed models for JA-based resistance, Thaler et al. (2004) found that jasmonate-deficient mutant tomato (*def1*) plants had no effect on susceptibility to the necrotrophic pathogen *Septoria lycopersici*, which showed the same percentage of necrotic leaf area compared to wild-type plants. Besides, the reduced jasmonate in *def1* plants resulted in increased susceptibility to the five intermediate pathogen species tested in a range of biotrophic and necrotrophic ends of the spectrum.

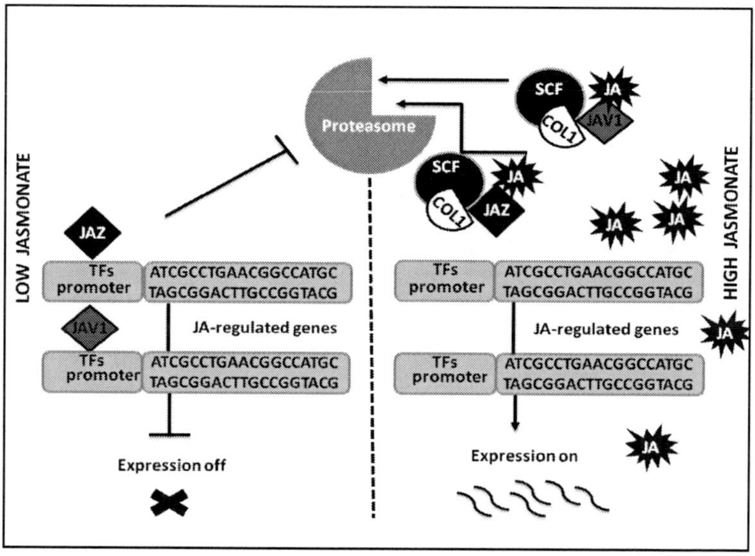

Figure 2. Jasmonic acid pathway regulators. Jasmonic acid (JA) in plant defense is controlled by the JAZ and JAV1 family of repressor proteins. In response to attackers, induced JA binds to the F-box protein (COI1 – coronatine insensitive1), a substrate-recruiting module of the Skp1-Cul1-F-box protein (SCF) ubiquitin E3 ligase complex to promote ubiquitination of JAZ and JAV1. Thus, degradation of both repressor proteins by the 26S proteasome derepress specific transcription factors and activate defense and developmental responses. Differently from JAZ, the JAV1 protein was recently identified as protein that allow the plant defense against necrotrophic pathogens and herbivorous insects apparently without influencing plant growth and development (Hu et al., 2013).

The study included two bacteria (*Pseudomonas syringae* and *Xanthomonas campestris*), two fungi (*Verticillium dahliae* and *Fusarium oxysporum* f. sp. *lycopersici*), and an oomycete (*Phytophthora infestans*). They concluded that when the jasmonate response is activated, even to the detriment of activation of the salicylate response, the plant will still induce resistance to taxonomically diverse pathogens with varying lifestyles and via different mechanisms.

Mur et al. (2006) also reported a synergistic interaction between SA and JA, in which tobacco plants with reduced SA levels, JA levels were not elevated during a hypersensitive response elicited by *Pseudomonas syringae* pv. *phaseolicola*. Truman et al. (2007) found that in systemically responding leaves challenged with an avirulent bacterial strain of *Pseudomonas syringae* pv. *tomato*, transcripts associated with JA biosynthesis, but not SA, were up-regulated in early stages. They proposed that JA and SA phases maybe act in tandem in the establishment of a broad systemic resistance. Zhu et al. (2014) demonstrated that both SA (MeSA) and JA (MeJA) are essential for systemic resistance response against TMV, possibly acting as the initiating signals for systemic resistance. The strongest systemic resistance against TMV was achieved after exogenous foliar application of JA followed by SA.

Additionally, Yi et al. (2014) studing the relationship between JA and SA signaling and bacterial flagellin (flg22) response in *A. thaliana* mutants showed that activated SA and JA signaling pathways had a positive influence on flg22-triggered oxidative burst, one of the early reactions triggered by flg22 perception.

2. JA-Mediated Defense Responses Requires Interconnected Hormonal Signal Transduction Pathways

Cross talk also occurs between JA/ABA, JA/ET, JA/AUX (auxin) and JA/CK (cytokinin) pathways, as well as other hormones, which have similarly optimized the adaptive defense response against pathogens and herbivores with different lifestyles (Koornneef and Pieterse, 2008; Meldau et al., 2012; Kazan and Lyons, 2014).

Genetic analysis have demonstrated that that the ABA signaling pathway interacts antagonistically with JA-ethylene (JA/ET) signaling pathway and vice versa to modulate plant development (Beaudoin et al., 2000; Ghassemian et al., 2000). Anderson et al. (2004) demonstrated a complex interplay between ABA and JA/ET signaling pathways in regulating plant defense gene expression and disease resistance. They showed that exogenous application of ABA suppressed both basal and JA-ethylene–activated transcription from defense genes. On the other hand, ABA deficiency mutants (*aba1* and *aba2*) resulted in the upregulation of basal and induced transcription from JA-ethylene responsive defense genes. Futhermore, they found that altered expression of AtMYC2 (a positive regulator of ABA signaling) transcripts during plant defense suggests a role for this gene in biotic stress in addition to its known role in ABA-mediated responses to abiotic stress.

Ethylene and jasmonate regulate a wide variety of physiological processes in plants, including the activation of specific responses to different types of stress, as well as plant defense (Lorenzo et al., 2003). This interaction can be both positive (O'Donnell et al., 1996) and negative (Rojo et al., 1999). In a synergistic cooperation, jasmonate and ethylene can induce pathogenesis related (PR) genes such as *PDF1.2* (Penninckx et al., 1998). It has been demonstrated that the constitutive expression of ETHYLENE RESPONSE FACTOR1 (ERF1), a downstream component of the ethylene signaling pathway (Solano et al., 1998), is an early ethylene- and jasmonate-responsive gene, and suggest that ERF1 may be a common component of both the ethylene and jasmonate signaling pathways (Lorenzo et al., 2003). The maize insect resistance 1-cysteine protease (Mir1-CP) is another gene dependent of ET and JA signaling pathways integration. Accumulation of Mir1-CP levels during fall armyworm (*Spodoptera frugiperda*) herbivory is increased by JA and ET phytormones. It was suggested that jasmonate act upstream of ethylene in the Mir1-CP transcription pathway, allowing for both an immediate response involving Mir1-CP accumulation and a delayed response inducing *mir1* transcript expression (Ankala et al., 2009). The clear interdependence of the ethylene and jasmonate pathways was demonstrated by Penninckx et al. (2013) in the induction of the plant defensin (*PDF1.2*) in Arabidopsis inoculated with *Alternaria brassicicola*.

A balance of both JA and ET compounds is pivotal to an effective defense response and their coordination is potentially specific in every plant, pathogen or insect species, as well as elicitor and target gene activated (Maffei et al. 2007; Pieterse and Dicke, 2007).

The interaction between JA and AUX is poorly understood, despite growing evidence of the AUX signaling involvement in both susceptibility and resistance to diverse pathogens (Llorente et al., 2008; Kazan and Lyons, 2014). Beneventi et al. (2013) proposed a plausible model for gene regulation during soybean-root knot nematode (RKN) incompatible interaction, in which endogenously elevated jasmonates stimulate auxin biosynthesis while auxin enhances ROS signaled induction. Additionaly, jasmonates can induce the accumulation of ascorbate, glutathione and increase the activity of dehydroascorbate oxidase involved in antioxidant activity.

Regarding to cytokinins (CKs), JA biosynthesis and JA-related defense metabolite accumulations are also correlated with this phytormone levels (Meldau et al., 2012). Several studies have showed that CKs enhance defense metabolite accumulation by regulating resource availability through increasing nitrogen and carbon allocation (Ruffel et al., 2011). Poplar upon wounding and CK treatments lead to higher transcripts levels of JA-biosynthetic genes and increased JA levels and its precursors (Dervinis et al., 2010).

In *A. thaliana* it has been shown that CK signaling might directly enhance JA-mediated defenses against pathogens (Argueso et al., 2009; Naseem et al., 2012). Ultimattely, the regulation of plastid-mediated defense responses such as the biosynthesis of JA, which is initiated in the chloroplast, may be dependent of CKs effect on plastid function and photosynthesis (Wasternack and Kombrink, 2010; Mitra and Baldwin, 2008).

Conclusion

A new scenario is emerging from complex hormonal crosstalk involved on plant disease and resistance regulation. Considerable progresses have been made in understanding the role of JA in plant defense mechanisms. The studies in model plants presented here show that the interaction of JA pathway with other hormones such as SA, ET, ABA, CK and AUX may result in a synergistic or antagonistic regulation during defense processes, depending on pathogen lifestyles and/or host genotype. The investigation of different resistant patterns exhibited by distinct plant-pathogen interactions provides an exciting research ground for the future.

The hormonal pathways may not only provide resistance against its proposed target group but also to species belonging to another group, such as necrotrophic and biotrophic pathogens. It was shown in tomato that the role of JA resistance response is involved in a wide range of plant attackers with diverse lifestyles, differently from the previous studies carried out with *A. thaliana* which suggested that distinct plant parasites are resisted via different mechanisms. Moreover, it has been suggested that the greater the overlap in the organisms that are affected by both JA and SA regulation, less unfavorable will be for the plants in their gene expression.

Finally, it is important to highlight that JA hormonal cross talk is not restricted for plant defense, but it may also be critical in plant development. A balance between the compounds of the interconnected pathways is necessary to not limit the plant growth and production. A better understanding of JA pathways during plant-pathogen interactions is needed to elucidate the molecules regulating such cross talk and several questions still remain to be answered around molecular mechanisms for the improvement of disease resistance.

REFERENCES

Anderson, J. P., Badruzsaufari, E., Schenk, P. M., et al. (2004). Antagonistic interaction between abscisic acid and jasmonate-ethylene signaling pathways modulates defense gene expression and disease resistance in arabidopsis. *The Plant Cell*, 16: 3460–3479.

Ankala, A., Luthe, D. S., Williams, W. P., and Wilkinson, J. R. (2009). Integration of ethylene and jasmonic acid signaling pathways in the expression of maize defense protein Mir1-CP. *MPMI*, 22, 1555–1564.

Argueso, C. T., Ferreira, F. J. and Kieber, J. J. (2009). Environmental perception avenues: the interaction of cytokinin and environmental response pathways. *Plant, Cell and Environ.*, 32: 1147–1160.

Balbi, V. and Devoto, A. (2008). Jasmonate signalling network in *Arabidopsis thaliana*: crucial regulatory nodes and new physiological scenarios. *New Phytologist*, 177: 301-318.

Beaudoin, N., Serizet, C., Gosti, F., and Giraudat, J. (2000). Interactions between abscisic acid and ethylene signaling cascades. *Plant Cell*, 12: 1103–1115.

Beneventi, M. A., Silva, O. B., Sá, M. E. L., et al. (2013). Transcription profile of soybean-root-knot nematode interaction reveals a key role of phythormones in the resistance reaction. *BMC Genomics*, 14: 1-17.

Carvalhais, L. C., Dennis, P. G., Badri, D. V., et al. (2013). activation of the jasmonic acid plant defence pathway alters the composition of rhizosphere bacterial communities. *PLoS ONE*, 8: 1-5.

Cheong, J. J. and Choi, Y. D. (2003). Methyl jasmonate as a vital substance in plants. *Trends Genet.*, 19:409-413.

Chini, A., Fonseca, S., Fernandez, G., et al. (2007). The JAZ family of repressors is the missing link in jasmonate signalling. *Nature*, 448:666–671.

Dalisay, R. R. and Kuc, J. A. Persistance of reduced penetration by *Colletotrichum lagenarium* into cucumber leaves with induced systemic resistqance and its relation to enhanced peroxidase and chitinase acitivities (1995). *Physiol. Mol. Plant Pathol.*, 47:329-338.

Delker, C., Stenzel, I., Hause, B., et al. (2006). Jasmonate Biosynthesis in *Arabidopsis thaliana – enzymes, products, regulation*. 8: 297-306.

Dervinis, C., Frost, C. J., Lawrence, S. D., et al. (2010). Cytokinin primes plant responses to wounding and reduces insect performance. *Journal of Plant Growth Regul.*, 29: 289–296.

Durrant, W. E. and Dong, X. (2004). Systemic acquired resistance. *Annu. Rev. Phytopathol.*, 42:185–209.

Felton, G. W., Korth, K. L., Bi, J. L., et al. (1999). Inverse relationship between systemic resistance of plants to microorganisms and to insect herbivory. *Cur. Biol.*, 9: 317–320.

Ghassemian, M., Nambara, E., Cutler, S., et al. (2000). Regulation of abscisic acid signaling by the ethylene response pathway in Arabidopsis. *Plant Cell*, 12: 1117–1126.

Glazebrook, J. (2005). Contrasting mechanisms of defense against biotrophic and necrotrophic pathogens. *Annu. Rev. Phytopathol.*, 43: 205–27.

Gupta, V., Willits, M. G. and Glazebrook, J. (2000). *Arabidopsis thaliana* EDS4 contributes to salicylic acid (SA)-dependent expression of defense responses: evidence for inhibition of jasmonic acid signaling by SA. *MPMI*, 13: 503–511.

Hammond-Kossack, K. and Jones, J. D. G. (2000). Responses to plant pathogens. In: *Biochemistry and Molecular Biology of Plants* (eds.) B. Buchanan et al. *American Society of Plant Physiologists*, 1102-1146.

Hammond-Kosack, F. and Jones, J. D. G. (1996). Resistance gene-dependent plant defense responses. *Plant Cell*, 8: 1773-1791.

Hu, P., Zhou, W., Cheng, Z., et al. (2013). JAV1 controls jasmonate-regulated plant defense. *Mol. Cell*, 50, 504–515.

Jones, J. D. and Dangl, J. L. (2006). Tha plant immune system. *Nature*, 444: 323-329.

Kazan, K. and Lyons, R. (2014). Intervention of phytohormone pathways by pathogen effectors. *The Plant Cell*, 26: 2285–2309.

Kessler, A. and Baldwin, I. T. (2001). Defensive Function of Herbivore-Induced Plant Volatile Emissions in Nature. *Science*, 291: 2141-2144.

Koornneef, A. and Pieterse, C. M. J. (2008). Cross talk in defense signaling. *Plant Physiol.*, 146:839-844.

Llorente, F., Muskett, P., Sánchez-Vallet, A., et al. (2008). Repression of the auxin response pathway increases Arabidopsis susceptibility to necrotrophic fungi. *Mol. Plant*, 1: 496–509.

Lorenzo, O. and Solano, R. (2005). Molecular players regulating the jasmonate signalling network. *Cur. Op. in Plant Biol.*, 8:532–540.

Lorenzo, O., Piqueras, R., Sánchez-Serrano, J. J., and Solano, R. (2003). Ethylene Response Factor 1 integrates signals from ethylene and jasmonate pathways in plant defense. *The Plant Cell*, 15: 165–178.

Maffei, M. E., Mithöfer, A. and Boland, W. (2007). Before gene expression: early events in plant insect interaction. *Trends Plant Sci.*, 12: 310-316.

McDowell, J. M. and Dangl, J. L. (2000). Signal transduction in the plant immune response. *Trends Biochem. Sci.*, 25:79-82.

Meldau, S., Erb, M. and Baldwin, I. T. (2012). Defence on demand: mechanisms behind optimal defence patterns. *Annals of Bot.*, 110: 1503-1514.

Mitra, S. and Baldwin, I. T. (2008). Independently silencing two photosynthetic proteins in *Nicotiana attenuata* has different effects on herbivore resistance. *Plant Physiol.*, 148: 1128–1138.

Mur, L. A. J., Kenton, P., Atzorn, R., Miersch, O., Wasternack, C. (2006). The Outcomes of Concentration-Specific Interactions between Salicylate and Jasmonate Signaling Include Synergy, Antagonism, and Oxidative Stress Leading to Cell Death. *Plant Physiol.*, 140(1): 249–262.

Naseem, M., Philippi, N., Hussain, A., et al. (2012). Integrated systems view on networking by hormones in Arabidopsis immunity reveals multiple crosstalk for cytokinin. *The Plant Cell*, 24:1793–1814.

O'Donnell, P. J., Calvert, C., Atzorn, R., Wasternack, C., Leyser, H. M. O., Bowles, D. J. (1996). Ethylene as a signal mediating the wound response of tomato plants. *Science* 274: 1914–1917.

Overmyer, K., Brosche´, M. and Kangasja¨, J. (2003). Reactive oxygen species and hormonal control of cell death. *Trends in Plant Sci.,* 8: 335-342.

Pauwels, L., Goossens, A. (2011) The JAZ proteins: a crucial interface in the jasmonate signaling cascade. *Plant Cell*, 23(9):3089-100.

Peer, R. V., Niemann, G. J. and Schippers, B. Induced resistance and phytoalexin accumulation in biological control of *Fusarium wilt* of carnation by *Pseudomonas* sp. Strain WCS417R. (1991). *Phytopathol.*, 81:728-734.

Penninckx, I. A. M. A., Thomma, B. P. H. J., Buchala, A., et al. (2013). Concomitant activation of jasmonate and ethylene response pathways is required for induction of a plant defensin gene in Arabidopsis. *The Plant Cell*, 10, 2103–2113.

Pieterse, C. M. J., Leon-Reyes, A., Van der Ent, S., and Van Wees, S. C. M. (2009). Networking by small-molecule hormones in plant immunity. *Nature Chem. Biol.*, 5: 308-316.

Pieterse, C. M. J. and Dicke, M. (2007). Plant interactions with microbes and insects: from molecular mechanisms to ecology. *Trends Plant Sci.*, 12: 564-569.

Pieterse, C. M. J., van Wees, S. C. M., van Pelt, J. A., et al. (1998). A novel signaling pathway controlling induced systemic resistance in Arabidopsis. *The Plant Cell*, 10: 1571–1580.

Pirbalouti, A. G., Sajjadi, S. E. and Parang, K. (2014). A review (research and patents) on jasmonic acid and its derivatives. *Arch. Pharm. Chem. Life Sci.*, 347: 229–239.

Robert-Seilaniantz, A., Grant, M. and Jones, J. D. G. (2011). Hormone crosstalk in plant disease and defense: more than just jasmonatesalicylate antagonism. *Annu. Rev. Phytopathol.*, 49: 317–43.

Rojo, E., León, J., Sanchex-Serrano, J. J. (1999). Cross-talk between wound-signalling pathways determines local versus systemic gene expression in *Arabidopsis thaliana*. *The Plant Journal*, 20,135–142.

Rossi, F. R., Gárriz, A., Marina, M., et al. (2011). The sesquiterpene botrydial produced by *Botrytis cinerea* induces the hypersensitive response on plant tissues and its action is modulated by salicylic acid and jasmonic acid signaling. *MPMI*, 24, 888–896.

Ruffel, S., Krouk, G., Ristova, D., et al. (2011). Nitrogen economics of root foraging: transitive closure of the nitrate-cytokinin relay and distinct systemic signaling for N supply vs. demand. *Proc. of the Nat. Ac.of Sci. of the US*, 108: 18524–18529.

Sheard, L. B., Tan, X., Mao, H., et al. (2010). Jasmonate perception by inositol phosphate-potentiated COI1-JAZ co-receptor. *Nature*, 468: 400–405.

Solano, R., Stepanova, A., Chao, Q. M., Ecker, J. R. (1998). Nuclear events in ethylene signaling: a transcriptional cascade mediated by *ETHYLENE-INSENSITIVE3* and *ETHYLENE-RESPONSE-FACTOR1*. *Genes and Development* 12:3703–3714.

Spoel, S. H., Koornneef, A., Claessens, S. M. C., et al. (2003). NPR1 modulates cross-talk between salicylate- and jasmonate-dependent defense pathways through a novel function in the cytosol. *The Plant Cell*, 15, 760–770.

Thaler, J. S., Owen, B. and Higgins, V. J. (2004). The role of the jasmonate response in plant susceptibility to diverse pathogens with a range of lifestyles. *Plant Physiol.*, 135: 530–538.

Thomma, B. P., Penninckx, I. A., Broekaert, W. F., and Cammue, B. P. (2001). The complexity of disease signaling in Arabidopsis. *Curr. Opin. Immunol.*, 13:63–68.

Truman, W., Bennett, M. H., Kubigsteltig, I., Turnbull, C., Grant, M. (2007). Arabidopsis systemic immunity uses conserved defense signaling pathways and is mediated by jasmonates. *Proc. Natl. Acad. Sci. US*, 104: 1075–1080.

Van Wees, S. C. M., Van der Ent, S. and Pieterse, C. M. J. (2008). Plant immune responses triggered by beneficial microbes. *Current Opinion in Plant Biol.*, 11:443–448.

Van Wees, S. C. M., de Swart, E. A. M., Van Pelt, J. A., et al. (2000). Enhancement of induced disease resistance by simultaneous activation of salicylate- and jasmonate-dependent defense pathways in *Arabidopsis thaliana*. *Proc. Natl. Acad. Sci. US*, 97: 8711-8716.

Wasternack, C. and Kombrink, E. (2010). Jasmonates: structural requirements for lipid-derived signals active in plant stress responses and development. *ACS Chem. Biol.*, 5: 63–77.

Wasternack, C. (2007). Jasmonates: an update on biosynthesis, signal transduction and action in plant stress response, growth and development. *Annals of Bot.*, 100: 681–697.

Worrall, D., Holroyd, G. H., Moore, J. P., et al. (2012). Treating seeds with activators of plant defence generates longlasting priming of resistance to pests and pathogens. *New Phytologist*, 193: 770–778.

Xu, Y., Chang, P. L., Liu, D., et al. (1994). Plant defense genes are synergistically Induced by ethylene and methyl jasmonate. *The Plant Cell*, 6, 1077-1085.

Yi, S. Y., Shirasu, K., Moon, J. S., et al. (2014) The activated SA and JA signaling pathways have an influence on flg22-triggered oxidative burst and callose deposition. *PLoS ONE*, 9: 1-10.

Zhu, F., Xi, D., Yuan, S., et al. (2014). Salicylic acid and jasmonic acid are essential for systemic resistance against tobacco mosaic virus in *Nicotiana benthamiana*. *MPMI*, 27: 567–577.

In: Jasmonic Acid
Editor: Lucille Morrison

ISBN: 978-1-63463-854-8

Chapter 2

SEPARATION AND DETECTION OF JASMONIC ACID IN PLANT SAMPLES

***Yehua Han*[1] *and Huwei Liu*[2*]**

[1]State Key Laboratory of Heavy Oil Processing, China University of Petroleum, Beijing, China

[2]Beijing National Laboratory for Molecular Sciences, Key Laboratory of Bioorganic Chemistry and Molecular Engineering of Ministry of Education, Institute of Analytical Chemistry, College of Chemistry and Molecular Engineering, Peking University, Beijing, China

ABSTRACT

Jasmonic acid (JA) is an essential phytohormone for plant development and stress response. It not only acts as the regulatory molecule in root elongation, germination, pollen production, fruit ripening and plant senescence, but also acts as the signaling compound activating plant defenses against pathogens, herbivory, wounding and abiotic stress. Determination of JA has been a crucial issue in plant research. Like all the other plant hormones, JA occurs at a trace level in a very complex matrix. So it is a challenging task to achieve the complete isolation and sensitive detection of JA. In the past few years, lots of researchers focus on this particular topic to develop various methods for JA separation and detection. The ultrasensitive and in-situ detection of JA

* Corresponding author. Tel.: +86 10 62754976; fax: +86 10 62751708. E-mail addresses: hwliu@pku.edu.cn.

has been achieved thanks to the latest advances in chromatography and mass spectrometry technologies. Nowadays even accurate analysis of JA stereoisomers has been achieved, offering a powerful tool to understand JA's stereospecific bioactivities. This paper mainly presents the recent development in separation and determination of phytohormone JA. According to diversified

1. INTRODUCTION

Plant hormones are naturally occurring organic substances that take part in the whole life cycle of plants at very low concentration level. According to diversified structures and physiological functions, plant hormones are generally grouped into several major classes including auxins, abscisic acid, cytokinins, gibberellins, ethylene, salicylic acid, jasmonates, brassinosteroids, nitric oxide, strigolactones, and polypeptide hormones. In this phytohormone kingdom, jasmonates are detected all through the plant and they play crucial roles to assist plants in adapting to abiotic or biotic changes by setting up the link between the environmental conditions and the intracellular response. Jasmonates act as the regulatory molecule in root elongation, germination, tuberization, tendril coiling, pollen production, fruit ripening and plant senescence [1, 2]. They also act as the signaling compound which activates plant defenses against pathogens, herbivory, wounding, ozone exposure and water deficit [2-4]. In jasmonates family, JA is no doubt the most prototypical member. JA is a small lipid that acts as a precursor for the production of jasmonate hormones [5] and all the other jasmonates members are JA's metabolites such as its methyl ester, glucose esters, hydroxylated forms, amino acid conjugates and other relative oxylipin derivatives.

1.1. Biosynthesis

JA was first isolated from culture filtrates of the fungus *Botryodiplodia* (syn. *Lasidiplodia*) theobromae in 1971 [6]. Decades later its biosynthetic pathway from linolenic acid was well established [7, 8]. The biosynthesis of JA starts with α-linolenic acid (18:3) originating from chloroplast membranes. Linolenic acid is oxygenated to form 13-hydroperoxy-9,11,15-octadecatrienoic acid (13-HPOT) under the catalyzation of 13-lipoxygenase (13-LOX). Then, 13-HPOT is oxidized to 13,13(S)-epoxy linolenic acid under the catalyzation of 13-allene oxide synthase (13-AOS). This epoxy compound is highly active and can be converted to *cis*-(+)-12-oxophytodienoic acid (*cis*-(+)-12-OPDA) under the catalyzation of allene oxide cyclase (AOC). *cis*-(+)-12-OPDA is reduced by saturation of the cyclopentenone double bond under the catalyzation of peroxisome OPR3 and subsequently β-oxidized by shortening of the carboxy side-chain under the catalyzation of enzymes specific for fatty acid β-oxidation to generate (+)-7-epi-JA. JA is to be further derivatized to form other jasmonates, for example, getting methyl jasmonate (MeJA) through methylation and getting JA-amino acid conjugates through conjugation.

1.2. Stereospecific Bioactivity

From the biosynthesis process of JA, it can be seen that the product is (+)-7-epi-JA whose stereostructure has been fixed as early as the production of *cis*-(+)-12-OPDA which is *cis*-orientated, dextrotatory and has the (9*S*,13*S*)-configuration. Although the *cis*-configuration of the side-chains is thermodynamically less stable compared with the *trans*-one, the *cis*-arrangement is maintained throughout these conversions *in vivo*. In fact, the general and enantiomeric structure is the chemical root for JA's specific bioactivities. Distinctive structural requirements for JA's activities were defined on the basis of various biological responses caused by JA treatment [9-12]. And the key points were summarized in the early stage [2, 10, 13], for example, an intact pentenyl side chain, a cyclopentanone ring and a keto group at C-6. A comparatively special observation is the stereoselectivity found in JA's bioactivity. As illustrated in the chemical structure (Figure 1), JA possessing two chiral centers at C-3 and C-7 can generate four possible stereoisomeric forms, (−)-(3R,7R)-JA (known as (−)-JA), (+)-(3S,7S)-JA ((+)-JA), (−)-(3S,7R)-JA ((−)-epi-JA) and (+)-(3R,7S)-JA ((+)-epi-JA), among which (−)-JA and (+)-JA are a pair of enantiomers in *trans* form while (−)-epi-JA and (+)-epi-JA are a pair of enantiomers in *cis* form. The epi-JAs are thermodynamically less stable due to higher steric hindrance between the *cis* side chains. So they have a tendency to epimerize at C-7 to their corresponding *trans* forms with the fixed stereochemistry at C-3 [12]. In other words, (−)-JA and (+)-epi-JA are considered as each other's epimers, and so are (+)-JA and (−)-epi-JA. In the plant kingdom, (−)-JA and (+)-epi-JA are believed to be the naturally existing stereoisomers [8] with (+)-epi-JA as the original one.

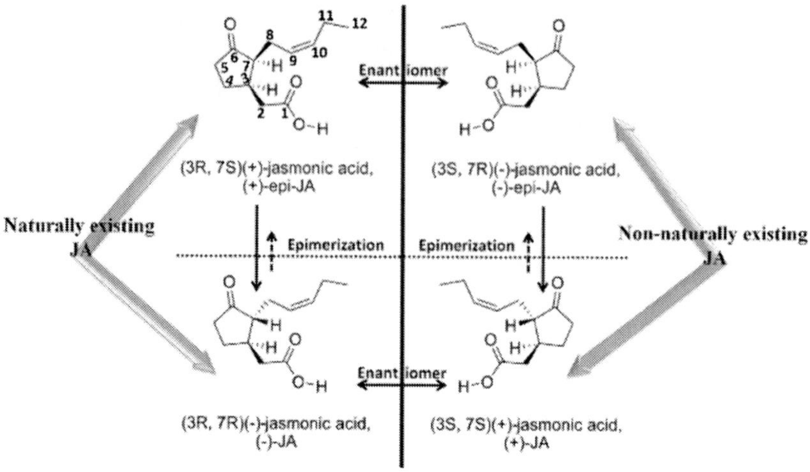

Figure 1. Chemical structures of four JA stereoisomers [55].

A better understanding of JA's regulation of plant growth and development requires accurate information about how JA naturally exists, such as the configuration, the amount and the distribution of JA when JA functions at the organ, cellular and sub-cellular level. Comprehensive determination of the JA is thus of great importance but still a difficult task. Because endogenous level of JA is quite low, usually at the ppb level in a complicated matrix together with other plant hormones, methods for thorough isolation of JA from limited plant materials, efficient separation of JA from other phytohormones and interferents as well as sensitive and selective detection of JA are highly needed to facilitate the investigation of JA functions. Besides, the stereoselective bioactivities of JA requires extremely efficient separation of JA's enantiomers or stereoisomers to provide detailed configuration information of JA and its related activities. In this review, recent advances in the techniques for extraction/purification, separation and detection of JA and related jasmonates compounds in plant samples are summarized and the corresponding contributions to the investigation of JA functions are highlighted.

2. State of the Art in JA Analysis

Techniques for JA analysis have never been stopped from advancing due to the fast development in analytical research field with the chromatography and mass spectrometry techniques as the most important contributors. Generally speaking, chromatographic methods make contributions partially in extraction/purification step and mainly in separation section, while mass spectrometry is the most powerful tool in detection stage. As well, other techniques are also adopted in JA analysis. For example, bioassays are classical methods for JA analysis but it has been attracted little attention in recent years because of tedious sample preparation, low specificity and poor repeatability. Meanwhile immunoassays have limited application considering the presence of cross reactions and the long period of antibody preparation despite of its specificity. Therefore, bioassays and immunoassays are not to be discussed in this review. Instead, chromatographic methods and mass spectrometry methods for JA determination are placed special importance and to be discussed in detail, along with some advancements in other techniques like electrochemical methods.

2.1. Extraction and Purification

Sample preparation is critical for almost all the subsequent assay steps, not only because JA exists at trace level but also because other abundant substances may interfere the final results if sample pretreatment is not properly processed. For the analysis of JA, sample preparation generally comprises several procedures, including sampling, freezedrying, comminution, homogenization, exhaustive extraction from the matrix, and subsequent removal of coextracted interferences by several cleanup steps prior to instrumental analysis. Solvent extraction is still the dominant method for JA extraction. The composition of solvent is the key factor to be optimized. A mixture of 1-propanol/H_2O/concentrated HCl [14-16], or methanol/water [17], or acetonitrile itself [18] has been proved to be able to extract JA. But none is speciallized for only JA, usually other plant hormone especially the acidic ones like gibberellic acid, indole-3-acetic acid, abscisic acid can be co-extracted. Crude extract from plants contains many interfering components so that direct analysis may not be feasible without further sample purification procedures.

When it comes to the purification step, three main measures are usually adopted: liquid–liquid extraction (LLE), solid phase extraction (SPE) and immunoextraction. Liquid-liquid extraction, also known as solvent extraction or liquid partition, is a method based on the solubility difference of target analytes between two liquid phases. Traditional LLE involves many extraction steps and thus is time-consuming and solvent-consuming. Thanks to the advancement of detection techniques, demand of sample amount is decreased and thus extraction solvent volume can be reduced accordingly , so LLE becomes easier to handle when only as little as 1 mL solvent is needed [14, 16]. A developed LLE method called liquid–liquid–liquid microextraction (LLLME) with good enrichment factors (48–243 fold) was applied to extract and concentrate JA and other three acidic phytohormones using hollow fibers [19, 20]. This LLLME method filled hollow fibers with phenetole as the organic phase, NaOH solution as the acceptor phase and HCl solution as the donor phase, utilized the pH difference between the donor and acceptor phases to preconcentrate analytes. Compared with LLE, solid-phase extraction is a sorbent based instead of solvent based sample pretreatment method. It utilizes the diverse interactions between compounds of interest and stationary phase to get controlled retention differences and to realize preseparation. SPE has benefits of low intrinsic costs, shorter processing times, low solvent consumption, and simpler processing procedures, and thus it is widely used in sample extraction and enrichment. Searching for suitable SPE columns, several factors should be taken into account: the sample matrix, analyte-specific physicochemical properties, and the nature of the bonded phase.The mainly used SPE cartridge for JA analysis is reversed phase C18 cartridge which is capable to get rid of lipids and other metabolites and retain JA under gradient elution[21]. Mixed mode SPE columns combining reversed-phase retention features with ion-exchange ability were also tested for JA pretreatment and MCX (strong cation exchanger) was proved to be the most suitable one among five commercially available solid-phase materials including MAX (strong anion exchanger), WAX (weak anion exchanger), MCX (strong cation exchanger), WCX (weak cation exchanger), and Oasis HLB (universal polymeric reversed-phase sorbent) [22, 23]. A miniaturized SPE method called solid-phase microextraction (SPME) replaces the column sorbent in SPE with fibers, making SPME a syringe-like device and allowing extraction, concentration, desorption and injection of JA to be carried out simultaneously [24]. Besides C18, other polymeric sorbent materials like poly (methacrylic acid-ethylene glycol dimethacrylate) monolith [25] or Super Q [15, 26] were also introduced for JA purification. Immunoextraction based on specific antibodies was achieved in the early age [27], but rarely used now because of cross-reactivity.

Generally, a typical procedure widely used for JA extraction and purification is still based on liquid extraction. The grounded plant material powder was added with internal standard and extracted by methanol/water followed by agitation and then centrifugation. After extraction and centrifugation, the supernatant was collected and evaporated under reduced pressure to remove methanol. The pH of the aqueous residue was adjusted to 2.8 with acetic acid. Then the acidic residue was partitioned twice against an equal volume of diethyl ether. After the aqueous phase was discarded, the organic fraction was vacuum evaporated at room temperature. The solid residue considered as relatively pure JA extract was re-suspended in methanol waiting for the next analysis.

2.2. Chromatographic and Mass Spectrometry Analysis

It is well known that, as for separation approaches, chromatographic techniques including liquid chromatography (LC), gas chromatography (GC) and capillary electrophoresis (CE) have been by now the most widely explored and useful tools for the high efficiency separation. As for detection approaches, mass spectrometry (MS) has been demonstrated to be especially powerful for sensitive, selective, qualitative and quantitative analysis. In the field of JA analysis, with analytes occurring at trace level and the requirement of simultaneously separation and detection with other phytohormones, chromatographic and mass spectrometry analysis no doubt play a crucial role.

GC-MS

GC method was first developed for MeJA determination which could be analyzed directly. For JA, its direct analysis by GC-MS is infeasible and derivatization to its methyl ester form (MeJA) is a necessary step, so that the analyte becomes less polar and more volatile, making it proper for GC or GC-MS analysis. Usually, JA's carboxyl group is methylated using ethereal diazomethane or boron trifluoride in methanol. After separation in GC, chemical ionization (CI) [15, 26, 28, 29] and electron ionization (EI) [30] are mainly used for MS detection of JA. CI is considered as a "soft" ionization method using lower energy, so that identifiable pseudomolecular ion $[MeJA+H]^+$ rather than fragment ions can be obtained in mass spectra, while EI yields more fragment ions giving a molecular fingerprint with structural information. After ionization, the measurements are performed by either selective ion monitoring (SIM) mode or full scan mode. GC-MS analysis of JA is well-developed and little innovation has been reported in recent years except for the sample preparation part designed for GC-MS. That is vapor phase extraction (VPE) technique [15, 26, 29]. It allowed one-step extraction, phase partitioning, methylation with HCl/methanol, and collection of methylated JA, and then volatilized compounds on Super Q filters, thereby omitting further purification steps and making this method highly efficient. Eluted samples were analyzed and quantified by GC-MS with good repeatability and linearity.

LC and LC-MS

Compared with GC-MS method, LC and LC-MS methods are not that routine and still under study. One advantage of LC and LC-MS methods for JA analysis is that no derivatization is needed and JA can be analyzed directly (using fluorescence detection is an exception). LC coupled with various types of detectors, including UV–vis, fluorescence detection, and especially MS, is the most practical method for JA determination.

Both UV and fluorescence detection are spectroscopic analysis, but the former is not so ideal for JA determination because JA has poor UV absorption and thus the detection sensitivity without preconcentration steps is relatively low [19]. An alternative means to the traditional UV method is fluorescence detection (FLD). Kristl et al. utilized 9-anthryldiazomethane to derivatize JA in situ with a detection limit of 2.9 ng/mL [31, 32]. The specificity and accuracy of the proposed LC/FD method was confirmed by liquid chromatography–electrospray ionization triple-quadrupole mass spectrometry analysis of free JA in *Lemna minor* samples under multiple reaction monitoring conditions. Chen et al. prepared a pre-column fluorescent derivatization reagent fluorescein-core piperazine (FCP) for JA labeling, giving a similar result as Kristl' work [33]. Xiong et al. developed an ultrasensitive and selective high-performance liquid chromatographic method based on precolumn derivatization of JA with 1,3,5,7-tetramethyl-8-aminozidedifluoroboradiaza-s-indacene (BODIPY-aminozide) [34]. The detection limit of the proposed method could reach as low as 1.14×10^{-10} M or 2.29 fmol per injection (20 μL).

LC-MS, in the other side, has drawn more and more researchers' interests and is becoming the most popular approach for JA analysis. Not only for analysis of JA individually, LC-MS has been applied for the simultaneous determination of JA and its metabolites or other multiple classes of plant hormones, facilitating the study of phytohormone function and crosstalk. One of the prominent advantages of LC-MS is that the accurate qualitative and quantitative analysis can be both achieved without many purification or derivatization steps. Almost all the LC-MS methods for JA analysis were developed in reversed-phase mode [16, 22, 23, 35-40], and all the separation was performed on various C18 columns. To distinguish JA from other phytohormones and their metabolites and to achieve simultaneous determination of them, gradient elution was adopted in the LC separation part. The mobile phase consisted of methanol or acetonitrile as the organic phase and water as the aqueous phase with formic acid or acetic acid as pH regulator or additive. Combined with LC, different kinds of MS detector could be chosen, including triple quadrupole (Q/Q/Q) [23, 35, 36, 39], time of flight (TOF) [37], quadrupole-trap (Q/trap) [16] and even high resolution MS orbitrap [40]. But as for the ionization source, electrospray ionization (ESI) was considered as the most suitable one because of JA's carboxyl group. The obtained limit of quantitation (LOQ) of all the reported LC-MS methods were in the range of 0.1-10 ppb. Different LC-MS methods were developed for different analyzing system with different phytohormones chosen as co-analytes using different plant materials. Theses co-analytes includes abscisic acid, salicylic acid, salicylic acid glucoside, tuberonic acid, tuberonic acid glucoside, indole-3-butanoic acid, cinnamic acid, benzoic acid, gibberellins, 13-epi-12-oxo-phytodienoic acid and zeatin. The improvement in either LC or MS technology has definitely promoted development of JA analysis. For example, the ultra-high performance liquid chromatography (UPLC) could provide high chromatographic resolution, high peak capacity and high speed analysis and thus a rapid and efficient separation of JA from its metabolites and other phytohormone classes [37, 38]. Besides, application of high resolution MS enable better discrimination of background and more accurate identification even without standard compounds [40].

CE and CE-MS

Among the high efficiency separation approaches, CE is recognized as a powerful microanalytical technique which integrates the advantages of both chromatographic methods and electrochemical methods. CE is versatile due to its multiple operation modes including capillary zone electrophoresis (CZE), micellar electrokinetic capillary chromatography (MEKC), microemulsion electrokinetic capillary chromatography (MEEKC), capillary gel electrophoresis (CGE), capillary isoelectric focusing (CIEF) and capillary isotachophoresis (CITP). Despite of its versatility, CE is not that widely used as LC because CE method is not that robust. However, it did contribute to some extent in the field of JA analysis [18, 41, 42]. Huang et al. developed a CE–UV method based on off-line membrane-supported liquid–liquid–liquid microextraction combined with on-column anion-selective exhaustive injection [41]. The proposed method was demonstrated to be suitable for the simultaneous quantification of JA and six other phytohormones with high sensitivity and good sample cleanup ability. Rather than UV, a special detector for CE is laser-induced fluorescence (LIF) detection whose sensitivity is higher than UV and even MS. Zhang et al. selected 5-bromomethylfluorescein as the derivatization reagent for JA to perform the CE-LIF analysis. This method enabled the analysis of JA in crude extracts without extra purification and enrichment procedures [42]. Another fluorescent reagent suitable for JA derivatization and the following CE-LIF analysis was 6-oxy-(acetypiperazine) fluorescein using 1-ethyl-3-(3-dimethylaminopropyl) carbodiimide as the condensing reagent [18]. MS, of course, can be a sensitive detector for CE. CE-MS method has been set up for JA analysis which is to be discussed in detail in the "Enantioselective analysis" section.

Other MS Methods

Separation of JA from other species is the guarantee to realize selective and sensitive determination of JA. Two ways could achieve this, either combining separation techniques like GC and LC before MS detection or improving the resolution of MS. For the latter, besides high resolution MS including TOF and orbitrap, differential mobility spectrometry–mass spectrometry (DMS–MS) was also proved to be a good choice [43]. DMS is an atmospheric pressure gas phase ion separation technique that operates as an ion filter between an ESI source and MS detector, introducing only preselected ions into the analytical area of MS. In DMS, ions are separated on the basis of mobility difference under high and low electric field. It offers continuous ion monitoring capability as well as orthogonal ion mobility separation so that high selectivity can be achieved [44-48]. Pang et al. developed a DMS–MS method for the direct separation and detection of five acidic phytohormones with JA included. Direct analysis of JA with MS was also realized by Bai et al. using direct analysis in real-time mass spectrometry (DART-MS) [20]. DART is an ambient ionization technique introduced by Cody et al. in 2005 [49]. DART-MS provides a rapid and straightforward way to perform the analysis, making itself a promising candidate as a screening approach. Its potential in endogenous JA analysis was demonstrated when coupled with single-drop liquid–liquid–liquid microextraction technique [20].

Enantioselective Analysis

An easily neglected but rather critical issue in the study of JA is the JA's stereospecific bioactivity. Usually, most of the analysis of JA did not consider JA as four stereoisomers when commercially available JA sample was applied or at least as two different epimers when endogenous JA was the target. However, the fact that there are four JA stereoisomers and only two of them including one thermodynamically less stable form are naturally existing is worthy of further study on the relationship between JA's biological properties and their stereospecificity. Different opinions were held on this hot topic [2, 13, 50-52]. Enantioseparation and selective detection of all the four JA stereoisomers is a must for the stereospecific bioactivity study. Chiral separation is a most challenging task in the separation field because the physical and chemical difference between a pair of enantiomers or epimers is much more difficult to tell than that between other kinds of isomers or different species of compounds. Chiral GC, chiral LC and chiral CE are all powerful tools in this field. For JA epimers separation and determination, GC equipped with permethylated β-CD coated column was first used for the chiral separation of JA [53]. Unfortunately, GC method required one to derivatize JA into MeJA first, so that epimerization possibly occurring during the derivatization process was out of control and direct and specific analysis of endogenous JA stereoisomers could not be achieved, besides, it was impossible to resolve JA and its methyl ester in one cycle separation and detection. Chiral CE and chiral LC methods have been reported by Han et al., giving a better solution to this problem [54, 55]. The reported CE-UV method based on short-end injection technique utilized α-CD as the chiral selector to realize direct resolution of both pairs of JA enantiomers within as short as 2.8 min [54]. A CE-MS method based on partial filling technique using the same chiral selector was developed accordingly to improve the detection sensitivity [54]. The reported normal phase LC-MS method utilized cellulose tris (4-methylbenzoate) coated silica gel as the chiral stationary phase to realize the simultaneous discrimination and direct analysis of all the four JA stereoisomers [55]. Endogenous rise of two natural JA stereoisomers was detected in tobacco leaves and their variations in response to mechanical wounding werc monitored. In addition, the configurational stability of JA stereoisomers was investigated using the stereoisomerically pure forms which were not commercially available but easily obtained by a related semi-preparative chiral LC method. Experimental evidence indicated that both of the two naturally existing JA stereoisomers were putative signals for wounding response, and the epimerization between them was not a spontaneous process simply promoted by the thermodynamical instability as expected before.

2.3. Electrochemical Methods

Electrochemical techniques, with benefits of simplicity, convenience and low cost, is another option for JA detection. Dang et al. reported two different electrochemical system for the determination of cis-jasmone, one was based on an electrodeposited Nafion film modified glassy carbon electrode in the presence of cetyltrimethylammonium bromide [56] and the other one was based on a multi-wall carbon nanotube/Nafion composite film modified glassy carbon electrode [57]. But the sensitivity of such kinds of methods could only reach ppm level, much lower than that obtained by chromatographic and mass spectrometry methods.

Summary and Prospects

Extensive studies on JA determination have developed various efficient analytical methods with different advantages. Among these, LC-MS methods, by now, have been proved to be the most powerful tool for JA separation and detection due to its superiority in sensitivity, accuracy, selectivity, reliability, simplicity, robustness and application. Direct analysis of endogenous JA as well as its stereoisomers in different plant samples has been achieved. However, an awkward situation is that all these developed methods including advanced instrumental techniques, separation strategies and detection approaches in analytical science have not been fully utilized by the plant biologists. Therefore many methods have been just proved the feasibility in JA determination rather than widely applied in practice to get more extensive information about how JA functions in diverse plant system. Though high cost of advanced instrument like UPLC-high resolution MS is one reason, the communication between analytical chemists and plant biologists is more dominant. Furthermore, to realize the direct analysis of JA in any plant samples which maybe rare and in very few amount, as well as to avoid hormone level changes occurring during the sample harvest process and pretreatment procedures, development of ultra-high sensitive (sub ppt level), in situ, transient, dynamic analytical methods for determining JA locally in a live plant accurately and precisely are especially attractive in the next stage of JA analysis.

REFERENCES

[1] Kessler, A., et al., Plant responses to insect herbivory: The emerging molecular analysis. *Annual Review of Plant Biology,* 2002. 53(2): p. 299-328.

[2] Wasternack, C., et al.,, Jasmonates: An update on biosynthesis, signal transduction and action in plant stress response, growth and development. *Annals of Botany,* 2007. 100(4): p. 681-697.

[3] Farmer, E.E., et al., Jasmonates and related oxylipins in plant responses to pathogenesis and herbivory. *Current Opinion in Plant Biology,* 2003. 6(4): p. 372-378.

[4] Yoshida, Y., et al., Jasmonic acid control of GLABRA3 links inducible defense and trichome patterning in Arabidopsis. *Development,* 2009. 136(6): p. 1039-1048.

[5] Farmer, E.E., et al., Plant biology - Jasmonate perception machines. *Nature,* 2007. 448(7154): p. 659-660.

[6] Beale, M.H., et al., Jasmonates: key players in the plant defence. *Natural product reports,* 1998. 15(6): p. 533-48.

[7] Chini, A., et al., The JAZ family of repressors is the missing link in jasmonate signalling. *Nature,* 2007. 448(7154): p. 666-U4.

[8] Vick, B.A., et al., Biosynthesis of jasmonic Acid by several plant species. *Plant physiology,* 1984. 75(2): p. 458-61.

[9] Yasunori Koda, Y.K., et al., Omparisons of various biological activities of stereoisomers of methyl jasmonate. *Phytochemistry,* 1992. 31(4): p. 1111-1114.

[10] Otto Miersch, et al., Structure-activity relations of substituted, deleted or stereospecifically altered jasmonic acid in gene expression of barley leaves. *Phytochemistry,* 1999. 50(3): p. 353-361.

[11] S. Blechert, C.B., et al., Structure-activity analyses reveal the existence of two separate groups of active octadecanoids in elicitation of the tendril-coiling response of Bryonia dioica Jacq. *Planta,* 1999. 207(3): p. 470-479.

[12] Holbrook, L.T., et al., Importance of the chiral centers of jasmonic acid in the responses of plants - Activities and antagonism between natural and synthetic analogs. *Plant Physiology,* 1997. 114(2): p. 419-428.

[13] Wasternack, C., et al., Jasmonate-signalled plant gene expression. *Trends in Plant Science,* 1997. 2(8): p. 302-307.

[14] Pan, X., et al., Quantitative analysis of major plant hormones in crude plant extracts by high-performance liquid chromatography-mass spectrometry. *Nature Protocols,* 2010. 5(6): p. 986-992.

[15] Schmelz, E.A., et al., Simultaneous analysis of phytohormones, phytotoxins, and volatile organic compounds in plants. *Proceedings of the National Academy of Sciences of the United States of America,* 2003. 100(18): p. 10552-10557.

[16] Pan, X., et al., Simultaneous quantification of major phytohormones and related compounds in crude plant extracts by liquid chromatography-electrospray tandem mass spectrometry. *Phytochemistry,* 2008. 69(8): p. 1773-1781.

[17] Chen, M.-L., et al., Highly sensitive profiling assay of acidic plant hormones using a novel mass probe by capillary electrophoresis-time of flight-mass spectrometry. *Journal of Chromatography B-Analytical Technologies in the Biomedical and Life Sciences,* 2011. 879(13-14): p. 938-944.

[18] Chen, H., et al., Simultaneous determination of phytohormones containing carboxyl in crude extracts of fruit samples based on chemical derivatization by capillary electrophoresis with laser-induced fluorescence detection. *Journal of Chromatography B-Analytical Technologies in the Biomedical and Life Sciences,* 2011. 879(20): p. 1802-1808.

[19] Wu, Y. et al., Simultaneous determination of several phytohormones in natural coconut juice by hollow fiber-based liquid-liquid-liquid microextraction-high performance liquid chromatography. *Journal of Chromatography A,* 2009. 1216(45): p. 7657-7663.

[20] Bai, Y., et al., Direct analysis in real time mass spectrometry combined with single-drop liquid-liquid-liquid microextraction for the rapid analysis of multiple phytohormones in fruit juice. *Analytical and Bioanalytical Chemistry*, 2012. 403(8): p. 2307-2314.

[21] Zhang, F.J., et al., Study on the Extraction, Purification and Quantification of Jasmonic Acid, Abscisic acid and Indole-3-acetic Acid in Plants. Phytochemical Analysis, 2008. 19(6): p. 560-567.

[22] Balcke, G.U., et al., An UPLC-MS/MS method for highly sensitive high-throughput analysis of phytohormones in plant tissues. *Plant Methods,* 2012. 8(1): p. 47-57.

[23] Bosco, R., et al., Development of an Ultrahigh-Performance Liquid Chromatography-Electrospray Ionization-Tandem Mass Spectrometry Method for the Simultaneous Determination of Salicylic Acid, Jasmonic

Acid, and Abscisic Acid in Rose Leaves. *Journal of Agricultural and Food Chemistry,* 2014. 62(27): p. 6278-6284.

[24] Zadra, C., et al., Quantification of jasmonic acid by SPME in tomato plants stressed by ozone. *Journal of Agricultural and Food Chemistry,* 2006. 54(25): p. 9317-9321.

[25] Zhang, W.P., et al., Polymer monolith microextraction moupled with HPLC for determination of jasmonates in wintersweet flowers. *Analytical Letters,* 2013. 46(1): p. 74-86.

[26] Schmelz, E.A., et al., The use of vapor phase extraction in metabolic profiling of phytohormones and other metabolites. *Plant Journal,* 2004. 39(5): p. 790-808.

[27] Albrecht, T., et al., Quantification of rapid, transient increases in jasmonic acid in wounded plants using a monoclonal antibody. *Planta,* 1993. 191(1): p. 86-94.

[28] Muller, A., et al., A multiplex GC-MS/MS technique for the sensitive and quantitative single-run analysis of acidic phytohormones and related compounds, and its application to Arabidopsis thaliana. *Planta,* 2002. 216(1): p. 44-56.

[29] Engelberth, J., et al., Simultaneous quantification of jasmonic acid and salicylic acid in plants by vapor-phase extraction and gas chromatography-chemical ionization-mass spectrometry. *Analytical Biochemistry,* 2003. 312(2): p. 242-250.

[30] Maksymiec, W., et al., The level of jasmonic acid in Arabidopsis thaliana and Phaseolus coccineus plants under heavy metal stress. *Journal of Plant Physiology,* 2005. 162(12): p. 1338-1346.

[31] Kristl, J., et al., Determination of jasmonic acid in Lemna minor (L.) by liquid chromatography with fluorescence detection. *Analytical and Bioanalytical Chemistry, 2005.* 383(5): p. 886-893.

[32] Kristl, J., et al., Evaluation of measurement uncertainty in the determination of jasmonic acid in Lemna minor L. by liquid chromatography with fluorescence detection. *Accreditation and Quality Assurance,* 2007. 12(6): p. 303-310.

[33] Chen, Y., et al., Fluorescein-core piperazine as a fluorescent labeling reagent for highly sensitive determination of jasmonic acid by high-performance liquid chromatography. *Analytical Methods,* 2013. 5(7): p. 1733-1738.

[34] Xiong, X.-J., et al., Ultrasensitive Determination of Jasmonic Acid in Plant Tissues Using High-Performance Liquid Chromatography with

Fluorescence Detection. *Journal of Agricultural and Food Chemistry,* 2012. 60(20): p. 5107-5111.

[35] Durgbanshi, A., et al., Simultaneous determination of multiple phytohormones in plant extracts by liquid chromatography-electrospray tandem mass spectrometry. *Journal of Agricultural and Food Chemistry,* 2005. 53(22): p. 8437-8442.

[36] Segarra, G., et al., Simultaneous quantitative LC-ESI-MS/MS analyses of salicylic acid and jasmonic acid in crude extracts of Cucumis sativus under biotic stress. *Phytochemistry,* 2006. 67(4): p. 395-401.

[37] Glauser, G., et al., Optimized liquid chromatography-mass spectrometry approach for the isolation of minor stress biomarkers in plant extracts and their identification by capillary nuclear magnetic resonance. *Journal of Chromatography A,* 2008. 1180(1-2): p. 90-98.

[38] Matsuura, H., et al., Simultaneous UPLC MS/MS analysis of endogenous jasmonic acid, salicylic acid, and their related compounds. *Plant Growth Regulation,* 2009. 57(3): p. 293-301.

[39] Mueller, M., et al., Rapid and sensitive hormonal profiling of complex plant samples by liquid chromatography coupled to electrospray ionization tandem mass spectrometry. *Plant Methods,* 2011. 7.

[40] Van Meulebroek, L., et al., Ultra-high performance liquid chromatography coupled to high resolution Orbitrap mass spectrometry for metabolomic profiling of the endogenous phytohormonal status of the tomato plant. *Journal of Chromatography A,* 2012. 1260(3): p. 67-80.

[41] Huang, L., et al., Membrane-supported liquid-liquid-liquid microextraction combined with anion-selective exhaustive injection capillary electrophoresis-ultraviolet detection for sensitive analysis of phytohormones. *Journal of Chromatography* A, 2014. 1343: p. 10-17.

[42] Zhang, Z.L., et al., Determination of jasmonic acid in bark extracts from Hevea brasiliensis by capillary electrophoresis with laser-induced fluorescence detection. *Analytical and Bioanalytical Chemistry*, 2005. 382(7): p. 1616-1619.

[43] Pang, N., et al., Study of field mobilities dependence and direct separation of acidic phytohormones by differential mobility spectrometry-mass spectrometry. *International Journal of Mass Spectrometry,* 2014. 362: p. 48-55.

[44] Krylov, E., et al., Field dependence of mobilities for gas-phase-protonated monomers and proton-bound dimers of ketones by planar

field asymmetric waveform ion mobility spectrometer (PFAIMS). *Journal of Physical Chemistry A,* 2002. 106(22): p. 5437-5444.
[45] Guevremont, R., High-field asymmetric waveform ion mobility spectrometry: A new tool for mass spectrometry. *Journal of Chromatography A,* 2004. 1058(1-2): p. 3-19.
[46] Kolakowski, B.M., et al., Review of applications of high-field asymmetric waveform ion mobility spectrometry (FAIMS) and differential mobility spectrometry (DMS). *Analyst,* 2007. 132(9): p. 842-864.
[47] Bohrer, B.C., et al., Biomolecule Analysis by Ion Mobility Spectrometry, in Annual Review of Analytical Chemistry. 2008. 5(3): p. 293-327.
[48] Kanu, A.B., et al., Ion mobility-mass spectrometry. *Journal of Mass Spectrometry,* 2008. 43(1): p. 1-22.
[49] Cody, R.B., Observation of Molecular Ions and Analysis of Nonpolar Compounds with the Direct Analysis in Real Time Ion Source. Analytical Chemistry, 2009. 81(3): p. 1101-1107.
[50] Fonseca, S., et al., The jasmonate pathway: the ligand, the receptor and the core signalling module. *Current Opinion in Plant Biology,* 2009. 12(5): p. 539-547.
[51] Schulze, B., et al., Rapid enzymatic isomerization of 12-oxophytodienoic acid in the gut of lepidopteran larvae. *Chembiochem,* 2007. 8(2): p. 208-216.
[52] Fonseca, S., et al., (+)-7-iso-Jasmonoyl-L-isoleucine is the endogenous bioactive jasmonate. *Nature Chemical Biology,* 2009. 5(5): p. 344-350.
[53] Schulze, B., et al., Profiling of structurally labile oxylipins in plants by in situ derivatization with pentafluorobenzyl hydroxylamine. *Analytical Biochemistry,* 2006. 348(2): p. 269-283.
[54] Han, Y., et al., Simultaneous discrimination of jasmonic acid stereoisomers by CE-QTOF-MS employing the partial filling technique. *Electrophoresis,* 2011. 32(19): p. 2693-2699.
[55] Han, Y., et al., Simultaneous determination of jasmonic acid epimers as phytohormones by chiral liquid chromatography-quadrupole time-of-flight mass spectrometry and their epimerization study. *Journal of Chromatography* A, 2012. 1235(6): p. 125-131.
[56] Dang, X.P., et al., A novel electrochemical method for determination of cis-jasmone based on enhancement effect of cetyltrimethylammonium bromide. *Journal of Electroanalytical Chemistry,* 2011. 657(1-2): p. 39-45.

[57] Dang, X., et al., Electrochemical characteristics of cis-jasmone in acid media at multi-wall carbon nanotube-Nafion composite film modified electrode and its analytical application. *Electrochimica Acta,* 2012. 81(1): p. 239-245.

In: Jasmonic Acid
Editor: Lucille Morrison
ISBN: 978-1-63463-854-8

Chapter 3

ROLE OF JASMONIC ACID IN INTERACTION OF PLANTS WITH PLANT GROWTH PROMOTING RHIZOBACTERIA DURING FUNGAL PATHOGENESIS

S. V. Veselova, T. V. Nuzhnaya and I. V. Maksimov
Institute of Biochemistry and Genetics,
Ufa Scientific Centre, Russian Academy of Sciences,
Bashkortostan, Russia

ABSTRACT

Jasmonic acid (JA) is a signal molecule of oxylipin nature, a classic participant of plant reactions to biotic stress coordinating their defense responses against necrotrophic pathogens and pests. The role of JA in regulation of induced systemic resistance (ISR), caused by beneficial microorganisms, the so-called plant growth-promoting rhizobacteria (PGPR), is well known. Development of ISR is manifested in such defense responses as deposition of callose, lignin, and phenolics at infection sites, increased activities of chitinase, peroxidase, polyphenol oxidase, and phenylalanine ammonia lyase (PAL), enhanced phytoalexin production, and induced expression of stress-related genes. This is likely to occur due to ability of PGPR to secrete a broad spectrum of metabolites, one of them being JA. Unfortunately, the mechanism of ISR regulation and the role of JA in it are not sufficiently studied.

Nevertheless, the importance of reactive oxygen species (ROS) in ISR induction under the influence of elicitors of bacterial nature is suggested.

The present report considers the aspects of the action of JA and entophytic bacteria *Bacillus subtilis* on the functioning of plant defense responses against fungus pathogens associated with activated transcription of PR-protein genes, intensity of accumulation of hydrogen peroxide, activation of peroxidase and lignin deposition in cell walls. It is shown that combined effects of endophytic strain of *B. subtilis* 26D and JA on defense response of wheat plants against hemibiotrophic and necrotrphic infection vary from stimulation to complete inhibition depending on concentration of signal molecule and succession of its application.

1. Introduction

In nature, plants are permanently threatened by a wide range of harmful pathogens and pests, including viruses, bacteria, fungi, oomycetes, nematodes and insect herbivores. Plant pathogens are mainly divided into biotrophs and necrotrophs according to their lifestyles. Biotrophs and necrotrophs are pathogens that derive nutrients from living host tissues and dead or dying cells, correspondingly. However, there are a lot of pathogens called hemibiotroph, which behave as either biotrophs or necrotrophs, depending on the stages of their life cycles, (Glazebrook 2005; Pieterse et al., 2009).

Plant defense reactions are usually complex and composed of several lines of constitutive (i.e. preformed) structural and inducible defense mechanisms that are effective against diverse array of pathogens. These inducible adaptive plant responses include multifaceted molecular, biochemical, and morphological changes, such as oxidative burst, expression of pathogenesis-related genes, production of antimicrobial compounds, and/or programmed cell death (Van Loon et al., 2006). The development of these induced responses of plants is coordinated by defense-related hormones (Bari and Jones 2009).

Three phytohormones— salicylic acid (SA), jasmonic acid (JA) and ethylene (ET), are known to play major roles in regulating plant defense responses against various pathogens and pests (Glazebrook 2005; Bari and Jones 2009; Ballare 2011; Pieterse et al., 2012). Classically, SA signaling triggers resistance against biotrophic and hemibiotrophic pathogens, whereas a combination of JA and ET signaling activates resistance against necrotrophic pathogens and pests (Glazebrook 2005; Robert-Seilaniantz et al., 2011).

Moreover, SA was initially associated with the induction of pathogenesis-related (PR) protcins and the establishment of systemic acquired resistance (SAR) (Ryals et al., 1996), whereas, JA is a key signal in the induced systemic resistance (ISR) caused by rhizosphere bacteria, the so-called plant growth-promoting rhizobacteria (PGPR) (Van Loon 2007; Vos et al., 2013).

JA and its derivatives are oxylipins, which are ubiquitous constituents in higher plants (Meyer et al., 1984). Jasmonic methyl ester was first isolated as odiferous constituent of the essential oil of *Jasminum grandiflorum* L. (Demole et al., 1962). JA accumulates rapidly in plants in response to wounding, herbivory and infection by necrotrophs. It was first connected with defense responses by the work of E.E. Farmer and C.A. Ryan on the regulation of expression of digestive proteinase inhibitors in tomato (*Solanum lycopersicum*) (Farmer and Ryan 1990; Farmer and Ryan 1992). Subsequently, the potential of jasmonates to protect plants from infection by pathogens has been shown for the interactions between potato and *Phytophthora infestans* (Cohen et al., 1993), between cotton and *Verticillium dahliae* (Li et al., 1996), and between *Arabidopsis* and *Pythium mastophorum* (Vijayan et al., 1998). In addition to its essential role in regulating disease and pest resistance, JA has also been implicated in senescence, root growth, fruit ripening, tendril coiling, pollen development, tuberization and responsiveness to abiotic stress (Ueda and Kato 1980; Cohen et al., 1993; Vos et al., 2013).

JA and its structurally related metabolites are synthesized in plants via the oxylipin biosynthesis pathway (Gfeller et al., 2010). JA biosynthesis starts with the release of α-linolenic acid (α-LA) from membrane lipids. Upon synthesis, JA can be readily metabolized to methyl jasmonate (MeJA) through the activity of JA CARBOXYL METHYLTRANSFERASE (JMT) (Seo et al., 2001) or conjugated to amino acids such as isoleucine via the activity of JA conjugate synthase (JAR1), which results in a biologically highly active enantiomer of jasmonoyl-isoleucine (JA-Ile) (Staswick and Tiryaki 2004).

The F-box protein CORONATINE INSENSITIVE1 (COI1) and transcriptional repressor proteins JASMONATE ZIM (JAZ) domain are key regulators of the JA signaling pathway. The perception of jasmonoyl-isoleucine (JA-Ile) is achieved by the ubiquitin ligase SCFCOI1 complex. When the F-box protein COI1 recognizes JA-Ile, it triggers the ubiquitination and subsequent proteosomal degradation of JAZ proteins. The degradation of JAZ proteins relieves JAZ-mediated repression of gene expression, leading to the activation of a large number of JA-responsive genes (Memelink 2009; Ballare 2011; Pieterse et al., 2012).

In *Arabidopsis*, there are two distinct branches of the JA signaling pathway: the MYC branch and the ERF branch. ERF-branch is associated with enhanced resistance to necrotrophic pathogens and is regulated by the APETALA2/ETHYLENE RESPONSE FACTOR (AP2/ERF)-domain containing transcription factors ERF1 and ORA59 (Vos et al., 2013). Activation of the ERF branch of the JA pathway requires both JA and ET signaling (Lorenzo et al., 2003; Pieterse et al., 2012). The MYC-branch is activated upon wounding and by attack of herbivorous insects and is regulated by the MYC transcription factors MYC2, 3 and 4 (Anderson et al., 2004; Pieterse et al., 2012). In addition, MYC2 also plays a role in priming for enhanced JA-responsive gene expression during PGPR-mediated ISR (Pozo et al., 2008).

JA along with ET plays a central role in the regulation of ISR. In Arabidopsis, the role of JA in the regulation of the ISR mediated by beneficial microorganisms was identified by analysis of the mutants with defects in JA-signaling such as *jar1*, *jin1* and *coi1* (Pieterse et al., 1998; Pozo et al., 2008). PGPR may activate inducible defense mechanisms in the plant in a similar way to pathogenic microorganisms. Such mechanisms can include reinforcement of plant cell walls, production of anti-microbial phytoalexins, synthesis of PR-proteins (Verhagen et al., 2004; Van Loon 2007). ISR-expressing plants display an accelerated defense response. PGPR is able to activate defense-related gene expression (Verhagen et al., 2004), by acting through 'priming' of the tissue for enhanced defense expression characterized by a faster and/or stronger activation of cellular defenses upon pathogen or insect attack resulting in enhanced resistance to the infection (Conrath et al., 2006; Van Loon 2007; Van Wees et al., 2008). Moreover the ISR response is also associated with accumulation of reactive oxygen species (ROS), callose deposition, and papillae formation at the site of pathogen entry (Van der Ent et al., 2009; Van Wees et al., 2008; Pozo et al., 2008). Besides, it is assumed that the early accumulation of ROS is one of the mechanisms triggering ISR response (Ahn et al., 2011).

This article is focused on some aspects of the action of JA and entophytic strain of *Bacillus subtilis* 26D, i.e. their capacity to activate transcription of JA-responsive genes, accumulation of ROS and the strengthening of the cell wall during pathogenesis.

2. THE ISR IN THE PLANT

Plant roots are surrounded by a nutrient-rich habitat, called the rhizosphere, which is the niche for a whole range of parasitic and beneficial microorganisms. Members of the genera *Bacillus*, *Erwinia*, *Pseudomonas*, *Rhizobium*, *Serratia*, and *Xanthomonas* are PGPRs with beneficial effects on plant performance (Van Loon 2007; Doornbos et al., 2011). They can promote plant growth, as well as suppress diseases. Plant growth promotion is believed to result from improved nutrient acquisition or hormonal stimulation (Van Loon 2007). The disease suppressive activity of PGPR is exerted either directly by hampering growth and development of soil-borne pathogens through competition for nutrients or secretion of antibiotics in the rhizosphere (Bakker et al., 2007; Van der Ent et al., 2009), or indirectly by eliciting a plant-mediated systemic defense response (Van Loon et al., 1998; Kloepper et al., 2004). This phenomenon, termed 'induced systemic resistance' (ISR) was first described by Van Peer et al., (1991) in carnation that was systemically protected against *Fusarium oxysporum* f.sp. *dianthi* upon treatment with strain *Pseudomonas fluorescens* WCS417, and by Wei et al., (1991) in cucumber, where six out of 94 rhizobacterial strains protected the leaves against anthracnose caused by *Colletotrichum orbiculare* (Van Peer et al., 1991; Wei et al., 1991).

PGPR-mediated ISR is effective against a broad-spectrum of plant pathogens, including oomycetes, fungi, bacteria and viruses (Van Loon et al., 1998), and even insect herbivores (Van Oosten et al., 2008) on many dicotyledonous plant species, including *Arabidopsis*, bean, carnation, eucalyptus, radish, tobacco, and tomato (Kloepper et al., 2004; Bakker et al., 2007; Van Wees et al., 2008), as well as monocots like rice, maize and wheat (Rosenblueth and Martines-Romero 2006; De Vleesschauwer et al., 2008). In addition, further studies revealed that endophytes (it is bacteria that colonize the internal tissues of plants) can be used as microbial inoculants to control plant pathogens and promote plant growth and yield, remove contaminants, solubilize phosphate, or provide available nitrogen to plants. (Rosenblueth and Martines-Romero 2006; Yi et al., 2013). Recent molecular studies on endophytic bacterial diversity have revealed a large richness of species, including members of the genera *Enterobacter, Pseudomonas, Serratia, Bacillus, Microbacterium, Streptomyces* (Rosenblueth and Martines-Romero 2006). For example, inoculation with endophytic strain *Bacillus pumilus* INR7 was effective against diseases caused by *Cucumber mosaic virus* (CMV),

Sclerotium rolfsii, *Ralstonia solanacearum*, *Colletotrichum gloeosporioides*, and *Rhizoctonia solani* in pepper and tomato (Yi 2013).

Induction of a plant-mediated ISR response starts with the recognition of the PGPR. It is well documented that PGPRs are specifically recognized by the plant through bacterial determinants referred to as microbe-associated molecular patterns (MAMPs). For PGPR, a lot of MAMPs have been identified, including flagellin and lipopolysaccharides (LPS), siderophores, antibiotics, biosurfactants, exopolysaccharides (EPS), and even volatile organic compounds (VOCs) shown to elicit ISR (De Vleesschauwer and Höfte 2009; Van der Ent et al., 2009; Verhagen et al., 2010; Fagar et al., 2013). Perception of MAMPs by pattern recognition receptors (PRRs) at the cell surface activates multiple host defense responses leading to a basal level of resistance (De Vleesschauwer and Höfte 2009).

In plants exhibiting ISR, a number of reactions have been observed such as deposition of callose, lignin, and phenolics at infection sites (Duijff et al., 1997), increases in stress-related enzyme activities such as phenylalanine ammonia-lyase (PAL), peroxidase, polyphenoloxidase, ß-1.3-glucanase and chitinase, enhanced phytoalexin production and induced expression of stress-related genes (Verhagen et al., 2004; Van Loon 2007; Verhagen et al., 2010), as well as enhanced capacity to express these defense responses upon challenge inoculation with a pathogen (Conrath et al., 2006).

Local and systemic defense responses triggered by beneficial and parasitic microorganisms are controlled by the plant hormones SA, JA and ET. There are many evidence that SA, JA, and ET pathway crosstalks allow plants to finely tune its defense responses depending on the type of pathogen (Pieterse et al., 2009; Pieterse et al., 2012). Pathogen-induced SAR is dependent on SA and the regulatory protein NPR1, as evidenced by the loss of SAR in transgenic NahG plants unable to accumulate SA, and in the *npr1* mutant (Sticher et al., 1997). The enhanced defensive capacity characteristic of SAR is always associated with the expression of PR-proteins. In contrast to SAR, rhizobacteria-mediated ISR is not associated with activation of the SA signalling pathway and with significant expression of known defense related genes, but requires responsiveness to JA and ET (Pieterse et al., 1998; Van Wees et al., 2008; Van der Ent et al., 2009; Pieterse et al., 2012). ISR seems to be based on increased sensitivity rather than on increased production of these hormones (Pieterse et al., 2000; Van der Ent et al., 2009). In accordance with the differences in signaling pathways, SAR and ISR were found to differ in their effectiveness against different types of attackers. Although ISR is active against the necrotrophic pathogens and pests, and SAR is effective against biotrophic pathogens, in some cases the spectrum of effectiveness of SAR and ISR is partly overlapping.

The role of JAs and ET in the regulation of PGPR-triggered systemic defense responses has been mainly established through the analysis of JA- and ET-signaling mutants. The use of the jasmonate response mutant *jar1*, the ET response mutant *etr1*, and the SAR regulatory mutant *npr1*demonstrated that signal transduction leading to *P. fluorescens* WCS417r–mediated ISR requires responsiveness to JA and ET and is dependent on NPR1 (Pieterse et al., 1998). Further studies have shown that transcripts of classic JA- and ET-regulated genes (e.g., *LOX1, LOX2, PAL1, HEL, CHI-B, PDF1.2*) in Arabidopsis plants expressing WCS417r-mediated ISR were not up-regulated, either before or after pathogen challenge (Van Wees et al., 1999). However, other research analysis of the *A. thaliana* transcriptome revealed that, locally in the roots, WCS417r bacteria inoculation stimulated a significant change in the expression of 97 genes without pathogen attack. Meanwhile no consistent alteration in gene expression was observed systemically in the leaves, demonstrating that the onset of ISR is not associated with detectable changes in gene expression. Nevertheless, after challenge inoculation with DC3000 of the plants pretreated with WCS417r bacteria, 81 genes showed a potentiated expression in the leaves, suggesting that these genes were primed to respond faster and/or more strongly upon pathogen attack (Verhagen et al., 2004).

Treatment with another rhizobacterium *Pseudomonas putida* strain LSW17S protected Arabidopsis from infection with *P. syringae* pv. *tomato* DC3000 through the typical phenomena of priming. Strong and faster transcription of defense-related genes *PR1*, *PR2*, *PR5*, and *PDF1.2* was induced in *Arabidopsis* treated with LSW17S and then infected with DC3000. In contrast, separate action of LSW17S and DC3000 did not elicit rapid molecular defense responses (Ahn et al., 2007). Similar to WCS417r - mediated ISR, LSW17S-mediated ISR depended on JA, ET and NPR1. Treatment with LSW17S reduced pathogen proliferation in *Arabidopsis* ecotype Col-0; however, in *npr1*, *etr1,* and *jar1* mutants the treatment with this bacterial strain did not inhibit pathogen growth. The *npr1*, *etr1*, and *jar1* mutations impaired the defense responses. Transcription of *PR1* and *PDF1.2* were almost completely abolished in the *npr1*, *etr1*, and *jar1* mutants treated with LSW17S and inoculated with DC3000 (Ahn et al., 2007).

Some specific strains of the species *Bacillus amyloliquefaciens, B. subtilis, B. pasteurii, B. cereus, B. pumilus, B. mycoides,* and *B. sphaericus* elicit ISR in a diversity of hosts and protect against various fungal and bacterial pathogens, viruses and nematodes. Studies on mechanisms indicate that elicitation of ISR by *Bacillus* spp. is associated with ultrastructural changes in plants during pathogen attack and with cytochemical alterations. Investigations into the signal transduction pathways of elicited plants suggest that *Bacillus* spp. activate the same pathways as *Pseudomonas* spp. and some additional pathways (Kloepper et al., 2004).

Priming provides a plant with an enhanced capacity for rapid and effective activation of cellular defense responses to effectively struggle against pathogen or insect attack. However, the molecular mechanisms underlying priming are still poorly understood. Hypothetically, the primed state is based on the accumulation, or posttranslational modification of one or more signaling proteins. Priming is expressed at the transcriptional level and transcription factor proteins are important members of the regulatory mechanism. Summarizing the literature data, it is possible to describe the following model for WCS417r-induced priming of JA-dependent defenses during ISR. The local onset of WCS417r-ISR in the roots requires responsiveness to ET (Knoester et al., 1999) and is associated with an ET-independent activation of the *MYB72* gene. Arabidopsis *myb72* mutants were incapable of mounting ISR against a set of (hemi)biotrophic and necrotrophic pathogens (Segarra et al., 2009; Van der Ent et al., 2008). The transcriptional regulator MYB72 is required in the roots during early signaling steps of rhizobacteria-mediated ISR, but not sufficient for the onset of ISR. Hence, MYB72 was assumed to act in concert with another signaling component. According to Van der Ent with co-workers (2008) ETHYLENE INSENSITIVE3 (EIN3)-LIKE3 transcription factor EIL3 is a potential candidate in this respect. The family of EIL proteins are key transcription factors of ET-regulated gene expression and act as positive regulators of ET signaling (Stepanova and Ecker 2000; Van der Ent et al., 2008).

Establishment of WCS417-ISR in systemic leaf tissues depends on the hormones JA and ET and requires the transcriptional (co)- activators NPR1 and MYC2 (Pozo et al., 2008; Van der Ent et al., 2009; Berendsen et al., 2012). The transcription factor MYC2 (originally called JIN1 for JASMONATE INSENSITIVE1) plays a central role in JA and abscisic acid-regulated signaling (Lorenzo and Solano 2005). MYC2-impaired *jin1* mutants were unable to mount WCS417r-ISR against *P. syringae* and *Hyaloperonospora parasitica* (Pozo et al., 2008) or *Piriformospora indica*-mediated ISR against *Golovinomyces orontii* (Stein et al., 2008), from which it follows that MYC2 is important regulator in priming during ISR. (Van der Ent et al., 2009). Moreover, WCS417r-ISR is associated with an increase in the level of *MYC2* mRNA (Pozo et al., 2008). In *npr1-1* mutant plants unable to mount an ISR response (Pieterse et al., 1998), the level of *MYC2* mRNA was not elevated upon colonization of the roots by WCS417r (Pozo et al., 2008). Thus, colonization of the roots by ISR-inducing WCS417r bacteria results in a moderate, but consistently enhanced expression of the transcription factor gene *MYC2*. In the JA signaling pathway, MYC2 functions as a positive regulator of JA-responsive genes such as *LOX2* and *VSP2*, (Lorenzo et al., 2004). In contrast, MYC2 acts as a negative regulator of JA-responsive genes such as *PDF1.2* and *HEL*, (Anderson et al., 2004; Lorenzo et al., 2004). Moreover, MYC2 modulates the expression of JA-responsive transcription factor genes, thereby affecting the expression of a large number of JA-responsive genes.

According to Pozo with co-workers (2008), colonization of the roots by ISR-inducing WCS417r bacteria leads to a systemic induction of the expression of the *MYC2* transcription factor gene. In the absence of a JA-inducing attacker, increased MYC2 levels do not result in enhanced defense-related gene expression, possibly due to suppression of MYC2 by JASMONATE-ZIM-DOMAIN (JAZ) repressor proteins of the JA pathway (Chini et al., 2007). Upon attack by a JA-inducing pathogen or insect, proteosomal degradation of JAZ proteins disables the JAZ-mediated suppression of the transcription factor activity of MYC2. This clears the way for the accelerated expression of MYC2-regulated JA-responsive genes, resulting in enhanced resistance against the JA-inducing attacker encountered (Pozo et al., 2008).

3. Role of Jasmonate Signaling in Regulation of ISR Mediated by PGPR

3.1. The Role of JA and Strain of *B. Subtilis* 26D in the Development of Resistance of Wheat Plants Infected with *S. Nodorum*

Generally, JA signaling activates resistance against necrotrophic pathogens, such as *P. mastophorum, Botrytis cinerea* and *Alternaria brassicicola* (Thomma 1998; Glazebrook 2005; Robert-Seilaniantz et al., 2011). However, there are publications confirming participation of JA signaling in the defense response against hemibiotrophic pathogens, such as *P. infestans* and *S. nodorum* (Cohen et a! 1993; Maksimov et al., 2014) and there is no evidence that JA-dependent responses protect against biotrophic pathogens, although this has not been investigated extensively (Glazebrook 2005). PGPRs are effective against a broad-spectrum of plant pathogens, including hemibiotrophic pathogens (Van Loon et al., 1998; Van der Ent et al., 2009; Yi et al., 2013). However, role of JA and PGPR in the resistance to hemibiotrophic pathogens are poorly understood, especially in wheat plants.

In our work we studied ascomycete *Septoria* (syn. *Stagonospora*) *nodorum* (Berk.) which is a major pathogen of wheat and causes leaf and glume blotch (septoria spot). This fungal pathogen belongs to hemibiotrophs which comprise a special sub-category of necrotrophs. The infection process of hemibiotrophs consists of asymptomatic period of proliferation of vegetative hyphae on the surface of the host with attempts to penetrate inside the host and the period of the destruction of the host cells, when hypha having penetrated through the epidermis of plant causes necrotic tissue damage (Oliver and Ipcho 2004).

To improve the efficacy of ISR, a combined application of inducing agents was employed. In many cases, a mixture of PGPR showed a more robust ISR response than did the single treatment (Raupach and Kloepper 1998; Jetiyanon and Kloepper 2002; Jetiyanon et al., 2003). By contrast, the efficacy of treatments with a combination of PGPR and chemical inducers is not well understood and was limited to treatment with SA or synthetic chemical inducers of SAR such as benzothiadiazole (BTH) (Van Wees et al., 2000; Anith et al., 2004; Yi et al., 2013). In our experiments, sequential treatment of wheat plants with JA in concentrations of 0.1 μM and 1 pM, and strain *B. subtilis* 26D resulted in a strong inhibition of disease symptoms, suggesting ISR against hemibiotrophic pathogen *S. nodorum* elicited by *B. subtilis* 26D strain and JA in wheat plants. Severe necrosis and chlorosis symptoms occurred in the control wheat seedlings, but were rarely observed in plants subjected to *B. subtilis* 26D or JA treatments (Table 1). However, co-treatment with *B. subtilis* 26D plus JA in two concentrations led to opposite results. We observed a clear additive effect of *B. subtilis* 26D + 1 pM JA treatment compared to treatment with *B. subtilis* 26D or JA alone (Table 1). Similar results have been shown for co-treatment with endophytic strain of *B. pumilus* INR7 and BTH on pepper plants (Yi et al., 2013). In contrast, *B. subtilis* 26D + 0.1 μM JA co-treatment resulted in a more severe chlorosis and necrosis compared to control (Table 1). In this case, we observed the effect of ISR suppression in wheat plants. Similar effect of defense response suppression was detected under co-application of SA and JA on Arabidopsis plants and it also depended on concentration of applied substances (Koornneef et al., 2008). Moreover, PGPR-mediated ISR is believed to enhance sensitivity of plant tissues to JA and other hormones (Pieterse et al., 2000; Van der Ent et al., 2009). This is a likely explanation of the additive effect detected in our case with lower JA concentration.

Table 1. Effects of JA and *B. subtilis* 26D on the development of septoria spot on the wheat seedling leaves 9 days after inoculation. Data are presented as means ± standard errors (n=9)

Treatments	Damage area, (mm^2)
Control	42.1±3.1
B. subtilis 26D	12.0±0.6
0.1 μM JA	15.3±0.8
B. subtilis 26D + 0.1 μM JA	38.6±1.6
1 pM JA	17.2±0.5
B. subtilis 26D + 1 pM JA	9.1±1.1

Elevation of defense related gene transcriptions has been assumed as molecular evidence of whether or not resistance is induced. Discrepancy between disease protection and transcription of defense-related genes has been an interesting theme in ISR. Defense priming is an important feature of ISR and it enhances capacity to express these defense responses upon challenge inoculation with a pathogen (Conrath et al., 2006).

To investigate the mechanisms of disease protection conferred by strain *B. subtilis* 26D, we examined the transcript accumulations of defense related genes, including PR genes, a *PR3* - chitinase I gene (*AB029936.1*), a *PR2* - β-1,3-glucanase gene (*DQ090946*) (Chen et al., 2008) and *PR-9* - lignin-forming peroxidase gene (*TC151917*) (Burkhanova et al., 2007) 24 and 72 h after of pathogen challenge. Sequential treatment of wheat plants with JA in two concentrations and *B. subtilis* 26D resulted in stronger and earlier increase in transcription activity of *PR-2*, *PR-3* and *PR-9* (Figure 1), compared to control, coinciding with increased plant resistance under these variants of treatment. Interestingly, the treatment of wheat plant with *B. subtilis* 26D strain did not result in significant changes in the transcription of the three defense genes without pathogen inoculation (Figure 1). In this case, we observed the typical phenomena of priming.

Thus rhizosphere bacterium *P. putida* LSW17S promoted strong and faster accumulation of transcripts of genes *PR-1*, *PR-2*, *PR-5* in Arabidopsis plants inoculated with the pathogenic bacterium *P. siringae* DC3000 (Ahn et al., 2007). In tomato plants inoculated with *Ralstonia solanacearum* after pretreatment with a cell-free filtrate of *Bacillus thuringiensis*, expression of defense-related genes such as *PR-1*, acidic chitinase, and β-1.3-glucanase was induced in stem and leaf tissues (Hyakumachi et al., 2013).

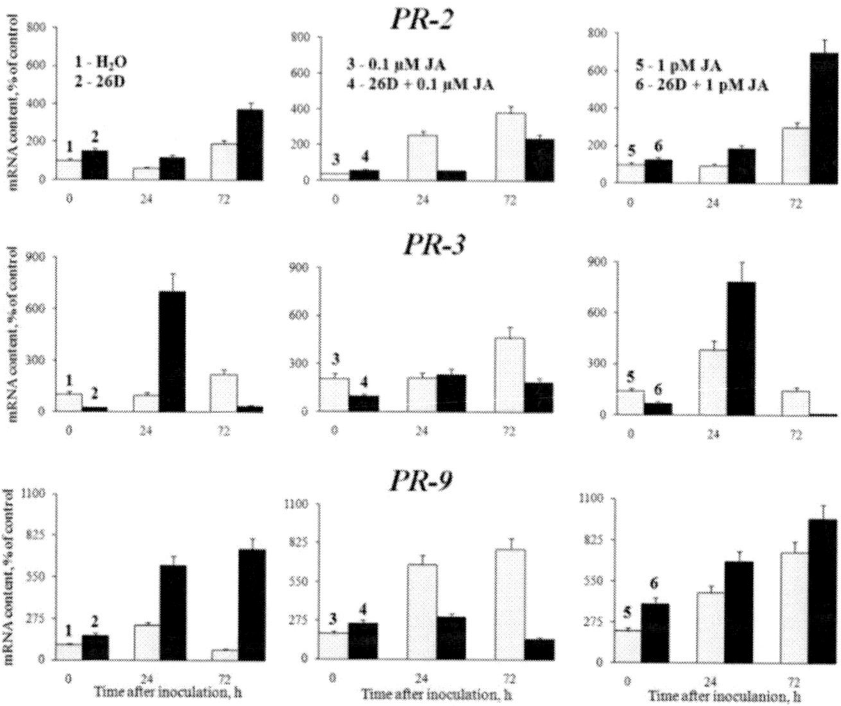

Figure 1. The effects of JA and *B. subtilis* 26D on relative mRNA content of defense related genes, including *PR2* - β-1,3-glucanase gene (*DQ090946*), *PR3* - chitinase I gene (*AB029936.1*) and *PR-9* - lignin-forming peroxidase gene (*TC151917*) in wheat leaves after infection with fungus *S. nodorum*. Data are presented as means ± standard errors (n=9).

The additive effect of *B. subtilis* 26D + 1 pM JA co-treatment was accompanied by a stronger up-regulation of all three genes compared to treatment with *B. subtilis* 26D or JA alone (Figure 1), whereas, the suppressive effect of *B. subtilis* 26D + 0.1 μM JA co-treatment was accompanied by a decrease in transcription of all three genes (Figure 1). Similarly, the endophytic strain *B. pumilus* INR7 and BTH combined treatment stimulated expression of pepper defense marker genes *CaPR1*, *CaTin1*, and *CaPR4* to a greater extent than did treatment with either agent alone (Yi et al., 2013). Data showing earlier accumulation of transcripts of the genes *PR-3* and *PR-9* in infected plants pretreated with *B. subtilis* 26D or *B. subtilis* 26D + 1 pM JA (Figure 1). Such an early activation of defense protein transcription was characteristic for plants treated with PGPR (Ahn et al., 2007).

It is well known that peroxidase, chitinase and β-1.3-glucanase are components of a broad-spectrum plant defense mechanism (Van Loon et al., 2006). Resistant tomato genotypes infected with *Alternaria solani*, showed high level of β-1.3-glucanase expression as compared to susceptible genotypes (Salim et al., 2011), while over-expression of rice chitinase gene *PR-3*(I) caused an increase in resistance to the agents of *R. solani* and *B. cinerea* infections in transgenic rice and cucumber plants (Lin et al., 1995; Tabei et al., 1998). Similarly, level of peroxidases was shown to be high in resistant compared to susceptible pea cultivars. (Chet and Inbar 1994).

PR-2, *PR-3* and *PR-9* expression is known to be regulated by JA (Wu and Bradford 2003; Kumari et al., 2006; Maksimov et al., 2014). Consequently, our results suggest that strain of *B. subtilis* 26D elicits mainly JA-dependent defense responses in wheat plants infected with *S. nodorum*.

3.2. The Effect of JA and *B. Subtilis* 26D on the Components of Pro- and Antioxidant Systems of Wheat Plants during the Defense Reactions against *S. Nodorum*

In addition to the production of PR-proteins the induced defense responses include hypersensitive response (HR), ion fluxes across the plasma membrane, production of ROS and reactive nitrogen species (RNS) (oxidative bursts), and reinforcement of the cell wall through lignification and deposition of callose (Almagro et al., 2009). A significant event in plant defense reactions is production of ROS, a common early response of host plant cells to pathogen attack and elicitor treatment. ROS play multifaceted functions in plants during defense responses including the oxidative burst capable of killing pathogen directly, establishment of physical barriers due to localized biosynthesis of lignin and suberin polymers, and phytoalexin production (Quan et al., 2008; Torres 2010). Furthermore, ROS also have a signaling function mediating defense gene activation and establishment of additional defenses by redox control of transcription factors (TFs) or by interaction with other signaling components (Kovtun et al., 2000; Mou et al., 2003; Torres 2010; Sewelam et al., 2013). Redox sensitive TFs activated by ROS then can stimulate the transcription of a large number of genes. Thus, expression of the 1.500 Arabidopsis transcription factors was monitored in response to different ROS, and ROS were found to alter the expression of about one-third of all known TFs in Arabidopsis (Gadjev et al., 2006). For example, members of genes family WRKYs and AP2/ERFs were highly responsive to ROS (Gadjev et al., 2006; Sewelam et al., 2013).

A strong interplay between ROS and other signaling molecules and phytohormones exists in HR and defense response, which explains the variety of signaling roles that ROS play under different conditions. ROS were proposed to be the central component that regulates interaction balance between SA, JA and ET to mediate some defense processes (Overmyer et al., 2003). DELLAs, transcription factors associated to gibberellin and phytohormone ABA can also regulate ROS production in response to pathogens (Achard et al., 2008; Kwak et al., 2003). Thus, hormones modulate disease resistance by interfering with signals transduction pathways mediated by SA, JA and ET (Torres 2010).

JA can indirectly influence accumulation of ROS. For example, JA regulates gene expression for polygalacturonase (PG), which activity correlates with accumulation of H_2O_2 in leaves under wounding. Oligogalacturonide fragments produced by PG are known to induce a substantial oxidative burst and the activation of defense gene expression (Orozco-Cárdenas et al., 2001). Another example is in that JA causes an increase in the activity of the NADPH oxidase and accumulation of ROS. It is assumed that by activating NADPH oxidase the exogenous elicitor leads to small local and transient accumulation of H_2O_2, which causes the synthesis of the JA. Then JA induces of large amounts of H_2O_2, which acts as a secondary messenger that induces the expression of the genes encoding the components of the JA biosynthes pathway (Hu et al., 2009).

Accumulation of ROS may play crucial role in the mechanism of ISR induced by nonpathogenic rhizobacteria. Plants treated PGPR show earlier production of ROS under inoculation of pathogens. This ability of bacteria is associated with the ISR-eliciting compounds, the so-called MAMPs such as cell wall LPS, siderophores, flagella, antibiotics, and volatile compounds, such as 2.3-butanediol, (Van Loon 2007; Van Wees et al., 2008). Protection of plants against pathogens with LPS from the nonpathogenic bacteria was associated with the oxidative burst concomitant with the production of ROS leading to induction of expression of several defense-related genes both locally and, to a much lesser extent, systemically (Van Loon et al., 2008).

In our work, sequential treatment of wheat plants with either JA strain *B. subtilis* 26D resulted in an increased generation of H_2O_2 6 h after inoculation of the *S. nodorum* in contrast to untreated control, in which oxidative burst was not observed (Figure 2). In case of *B. subtilis* 26D + 1 pM JA co-treatment, we observed an additive effect on H_2O_2 accumulation at the same time (Figure 2). The rapid accumulation of ROS in these variants of treatments was consistent with defense gene expression and reduced disease symptoms (Table 1, Figure 1). By contrast, plants pre-treated with *B. subtilis* 26D + 0.1 μM JA did not show any accumulation of H_2O_2 6 h after inoculation and no increase in the expression of defense-related genes (Figure 1, Figure 2) resulting in severe disease symptoms (Table 1).

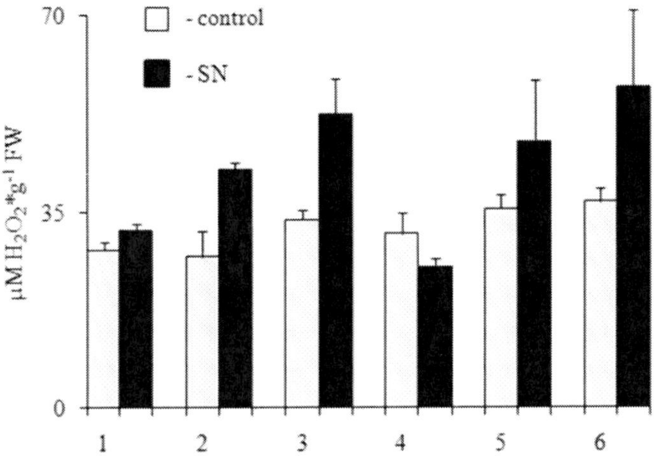

Figure 2. The effects of JA and *B. subtilis* 26D on the generation of H_2O_2 in the wheat leaves 6 h after inoculation of the *S. nodorum* (SN): 1 - nontreated control; 2 – *B. subtilis* 26D; 3 - 0.1 μM JA; 4 - *B. subtilis* 26D + 0.1 μM JA; 5 - 1 pM JA; 6 – *B. subtilis* 26D + 1 pM JA. Data are presented as means ± standard errors (n=9).

Similarly, the rhizobacterium, *P. putida* strain LSW17S, protected tomato plants from bacterial speck by biotrophic *P. syringae* pv. *tomato* strain DC3000 through the typical phenomena of priming. DC3000 challenge following LSW17S triggered rapid transcriptions of PR genes and H_2O_2 production in leaves and roots. H_2O_2 production is one of the earliest responses induced by DC3000 infection in the primed plants. According to this interpretation, earlier induction of ROS and transcription of defense-related genes were related with the ISR or priming effective for inhibition of DC3000 infection (Ahn et al., 2011).

For example, higher levels of H_2O_2 content were correlated with reduced anthracnose rot in methyl jasmonate-treated loquat fruit (Cao et al., 2008), and the lower *Rhizopus* rot caused by *R. stolonifer* in peach fruit treated with the biocontrol agent *B. cereus* AR156 and *B. subtilis* SM21 was closely related to increased H_2O_2 content, markedly up-regulated expression of chitinase, β-1.3-glucanase and PAL and promotion of their enzyme activities (Wang et al., 2013). Our results suggest, that H_2O_2 should act as a signaling molecule against *S. nodorum*, and that nonpathogenic bacterial strain *B. subtilis* 26D triggers ISR in wheat plants dependent on JA through the mechanism of accumulation of H_2O_2, leading to the expression of JA-responsible defense-related genes, which corresponds to literature data (Ahn et al., 2007; Van Loon et al., 2008; Verhagen et al., 2010; Ahn et al., 2011; Wang et al., 2013).

In plants infected with pathogens, NADPH oxidases located on the external face of the plasma membrane and peroxidases, polyamine oxidases and oxalate oxidases located in the cell wall matrix are the sources of ROS. Unlike polyamine oxidases and oxalate oxidases, which directly generate H_2O_2, both NADPH oxidases and peroxidases catalyse the initial formation of superoxide anion, which later dismutates to H_2O_2, which is the most stable species of ROS (Almagro et al., 2009). Peroxidases are a well-known class of PR proteins, and are induced in host plant tissues by pathogen infection. They belong to the PR-protein 9 subfamily (van Loon et al., 2006), and are expressed to limit cellular spreading of infection through establishment of structural barriers or generation of highly toxic environment by massive production of ROS and RNS (Passardi et al., 2005; Maksimov et al., 2011b). Peroxidases are soluble apoplastic and cell wall-bound isoenzymes. Probably due to high number of their isoforms, and due to the heterogeneous regulation of their expression, peroxidases are involved in numerous cellular processes during plant development and stress response (Passardi et al., 2005; Kumari et al., 2006). Expression of peroxidase genes in higher plants is induced by various stimuli such as pathogen infection, wounding, oxidative stress and phytohormones. JA and MeJA positively regulate some peroxidases genes expression (Repka et al., 2004; Ali et al., 2006; Kumari et al., 2006).

Although the key function of plant peroxidases is in oxidation of phenolic substrates at the expense of H_2O_2, peroxidases may, paradoxically and transitorily, generate H_2O_2 (Almagro et al., 2009). In our research, significant increase in peroxidase activity was detected in parallel with H_2O_2 generation in infected plants pre-treated with *B. subtilis* 26D or JA alone, and with mix of *B. subtilis* 26D + 1 pM JA (Table 2). Suppressive effect of *B. subtilis* 26D + 0.1 μM JA co-treatment was expressed in low peroxidase activity and significant drop of H_2O_2 content (Table 2), resulting in severe disease symptoms (Table 1). Similarly, pretreatment with JA of plants infected with *Ralstonia solanacearum* was shown to result in sharp increase in peroxidase activity peaking at 24 h (Mandal et al., 2013). Peroxidase activity also increased in millet (Deepak et al., 2007), pea (Kumari et al., 2006), arabidopsis (Traw et al., 2003), potato (Maksimov et al., 2014) and plants of many other species treated with JA. Bacteria, including those of the genus *Bacillus*, are also known to increase peroxidase activity, while inducing basal resistance to pathogens (Bargabus et al., 2002; Nakkeeran et al., 2006; Wang et al., 2013; Maksimov et al., 2014). Thus it has been shown recently that pretreatment with strain *B. subtilis* 26D, led to a sharp increase in peroxidase activity in potato plants infected with *P. infestans* (Mont.) (Maksimov et al., 2014). In plants, peroxidase activity develops due to synthesis and activity of numerous iso-enzymes, coded by multi-gene family, which functions are practically not defined until now (Cosio and Dunand 2009). Our results indicate that significant changes in activity of peroxidase in infected wheat is due to induced transcription activity of the gene coding for lignin-forming peroxidase (*TC151917*) (Figure 1), capable of binding to chitin and fungus mycelium (Maksimov et al., 2005) and participating in lignin deposition (Hiraga et al., 2001; Maksimov et al., 2014).

Table 2. Effects of JA and *B. subtilis* 26D on generation of hydrogene peroxide and activity of peroxidase in wheat leaves under *S. nodorum* infection. Data are presented as means ± standard errors (n=9)

Treatments	H_2O_2, (μM/g FW)		Peroxidase activity, (OD/mg protein)	
	Time after inoculation, h			
	24 h	48 h	24 h	48 h
Healthy plants	20.0±0.6	28.7±1.1	25.5±2.3	22.7±2.2
Septoria nodorum (SN)	24.7±1.2	15.2±0.4	40.0±4.1	29.3±3.3
B. subtilis 26D	30.8±1.1	22.9±1	31.5±2.5	39.1±2.1
B. subtilis 26D + SN	**40.1±1.8**	37.1±1.2	48.8±2.6	**50.1±5.3**
0.1 μM JA	33.3±2.2	20.1±0.8	41.2±3.3	45.3±4.4
0.1 μM JA + SN	**50.5±2.0**	38.3±1.5	**61.2±6.7**	**75.1±4.7**
B. subtilis 26D + 0.1 μM JA	29.5±0.3	17.7±0.6	26.2±2.8	19.7±2.1
B. subtilis 26D + 0.1 μM JA + SN	**13.6±0.2**	**4.1±0.2**	**25.3±2.8**	**13.4±0.1**
1 pM JA	27.0±3.0	29.5±0.8	45.3±4.8	48.2±2.5
1 pM JA + SN	39.2±1.9	23.4±1.3	54.4±3.3	54.0±0.9
B. subtilis 26D + 1 pM JA	39.6±1.2	23.3±1.4	74.2±3.9	57.8±3.0
B. subtilis 26D + 1 pM JA + SN	**55.3±2.8**	39.6±1.6	**81.1±4.2**	**82.6±3.2**

We suggest that it is parallel effect of H_2O_2 accumulation and peroxidase activation that result in subsequent development of resistance to septoria spot.

It is known that the peroxidase can create a physical barrier to limit pathogen invasion in host tissues by catalyzing the deposition of lignin and suberin (Ros Barcelo 2007). Both lignification and suberization involve formation of a three-dimensional poly(phenolic) matrix, initially within the carbohydrate matrix of the primary cell wall (Keren-Keiserman et al., 2004). Lignification and suberization are H_2O_2-dependent peroxidase-mediated processes, in which H_2O_2 is necessary for the peroxidative oxidation of monolignols and phenolics components (Vanholv et al., 2010; Miedes et al., 2014). The lignin biosynthetic pathway, where one of the main enzymes is PAL is generally divided into two branches: the general phenylpropanoid pathway and the monolignol-specific pathway. After their biosynthesis, the monolignols are transported to the cell wall where they are oxidized by peroxidases to monolignol radicals (Vanholv et al., 2010; Miedes et al., 2014). Phenylpropanoid pathway is required for the synthesis of numerous other phenolic compounds such as phenylpropanoid conjugates, flavonoids and phytoalexins (Miedes et al., 2014), which synthesis also involves peroxidases (Okazaki et al., 2007).

Evidence for a role for lignin and soluble phenolics in plant defense has been obtained from the analysis of the pathogen resistance of transgenic plants and mutants with contrasting lignin amount or composition (Miedes et al., 2014). For example, transgenic tobacco plants constitutively overexpressing *PAL* genes showed a higher tolerance toward *Cercospora nicotianae* and *Phytophthora parasitica* pv. *nicotianae* (Way et al., 2002; Way et al., 2011; Shadle et al., 2003). In cotton (*Gossypium hirsutum*), quantitative analysis of resistance to the wilt fungus *Verticillium dahliae* revealed an association between increased lignification in the stems upon infection and resistance against wilt (Xu et al., 2011). In melon (*Cucumis melo*), lignin accumulation upon infection was found to increase faster and to a higher level in lines resistant to the powdery mildew fungus *Podosphaera fusca* than in susceptible lines, and this differential accumulation correlated with enhanced PAL levels (Romero et al., 2008).In wheat *(Triticum monococcum)*, silencing the monolignol biosynthesis genes *TmPAL*, *TmCOMT*, *TmCCoAOMT*, and *TmCAD* led to super-susceptibility of leaf tissues to the fungus *Blumeria graminis* f. sp. *tritici*, the causal agent of powdery mildew disease (Bhuiyan et al., 2009).

In our work, treatment of wheat plants with either JA or *B. subtilis* 26D or both *(B. subtilis* 26D + 1 pM JA) resulted in increased deposition of lignin 72 h after inoculation of the *S. nodorum* compared to untreated control, in which deposition of lignin was scarcely observed (Figure 3).

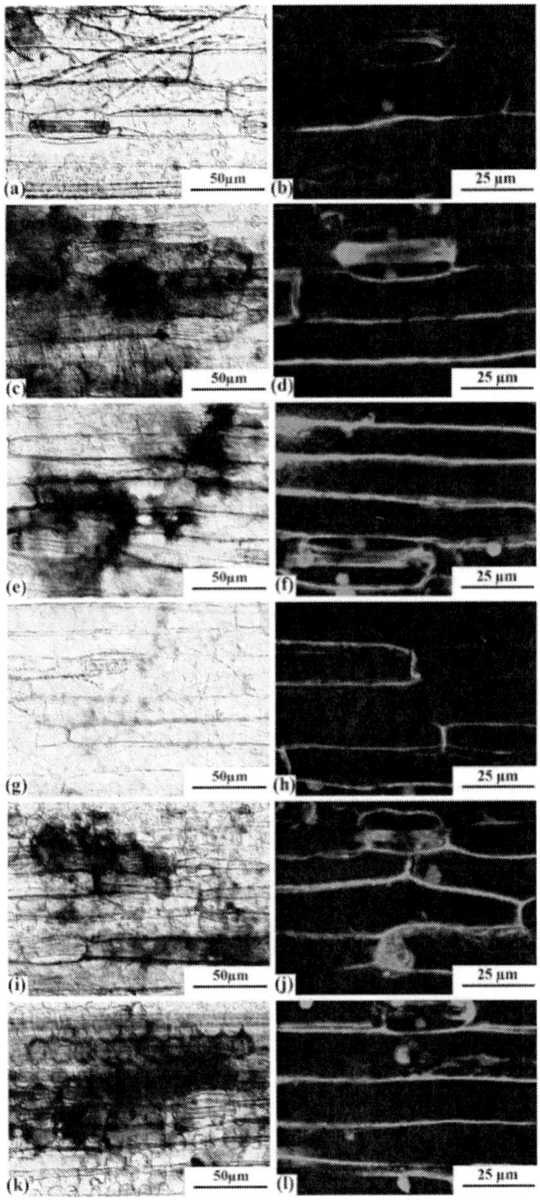

Figure 3. The effects of JA and *B. subtilis* 26D on localization of H_2O_2 (a, c, e, g, i, k) and deposition of lignin-like phenol compound (b, d, f, h, j, l) in the wheat leaves 24 h and 72 h after infections with fungus *S. nodorum* correspondingly: a, b - nontreated control; c, d – *B. subtilis* 26D; e, f - 0.1 μM JA; g, h - *B. subtilis* 26D + 0.1 μM JA; i, j - 1 pM JA; k, l – *B. subtilis* 26D + 1 pM JA.

It is of interest that lignin deposition detected at the sites of pathogen penetration correlated with increased generation of H_2O_2 in the same sites (Figure 3) measured with the help of vital staining of leaf with 3.3-diaminobenzidine (De Vleesschauwer et al., 2010). By contrast, plants pre-treated with *B. subtilis* 26D + 0.1 μM JA did not show any sufficiently deposition of lignin 72 h after inoculation and no accumulation of H_2O_2 at the sites of pathogen penetration (Figure 3), resulting in severe disease symptoms (Table 1).

Unfortunately, little is known about the ability of these PGPR strains to stimulate higher levels of defense enzymes, such as PAL, peroxidase and polyphenol oxidase (PPO). PAL plays an important role in the biosynthesis of phenolics and phytoalexins. The increase in PAL activity indicates activation of the phenylpropanoid pathway. Peroxidase and PPO catalyze the last step in the biosynthesis of lignin and other oxidized phenols. *Pseudomonas aeruginosa* 7NSK2 activated PAL in bean roots (De Meyer et al., 1999) and *B. subtilis* AF1 resulted in a distinct increase in the activity of PAL in pigeon pea (Podile and Laxmi 1998). *B. subtilis* strain BSCBE4 and *P. chlororaphis* strain PA23 increased activity of PAL, peroxidase and PPO in hot pepper seedlings challenged with *Pythium aphanidermatum* (Nakkeeran et al., 2006). Higher peroxidase activity was observed in cucumber roots treated with *P. corrugata* challenged with *P. aphanidermatum* (Chen et al., 2000).

It is known that, peroxidase genes expression (Repka et al., 2004; Ali et al., 2006; Kumari et al., 2006) and the phenylpropanoids pathway (Alon et al., 2013) are positively regulated by JA. JA treatment induced expression of six peroxidase genes in rice plants, besides, JA-inducible peroxidases genes responded to infection with rice blast fungus (Sasaki et al., 2004). JA treatment induced a significant increase in the activities of PAL, peroxidase, PPO and enhanced the content of phenolic compounds and deposition of lignin in tomato plants inoculated with *R. solanacearum* (Mandal et al., 2013). Treatment of Arabidopsis seedling with JA caused extensive lignifications (Ellis et al., 2002; Cano Delgado et al., 2003; Hamann et al., 2009; Denness et al., 2011).

Unfortunately, the mechanism of lignin production in the plant-pathogen system is insufficiently studied. The enzymes of some pathogens, including pectinase, protease and cellulase degrade the cell wall of the host plant and release nutrients for pathogen growth; thereby they induce synthesis of JA and subsequent defense responses with the induction of the strengthening of the plant cell wall (Cano Delgado et al., 2003). JA is assumed to inhibit cellulose synthesis and to lead to accumulation of lignin (Cano Delgado et al., 2003).

Recently two stages and several components of a genetic network have been identified that regulate cell wall damage (CWD)-induced lignin production in Arabidopsis. During the early stage, calcium and ROS production are required to induce a secondary ROS burst and JA accumulation. During the second stage, ROS derived with the help of NADPH oxidase and JA-isoleucine generated by JASMONIC ACID RESISTANT1 (JAR1), form a negative feedback. This feedback loop in turn seems to influence lignin accumulation (Denness et al., 2011).

Thus, PGPRs induce ISR by activating protective enzymes either through their gene expression or activity of proteins themselves thereby changing ultra-structure of plant cells. This is due to the early accumulation of ROS which may play crucial role in the mechanism of ISR induced by PGPR. Besides, it is assumed that the early accumulation of ROS is one of the mechanisms triggering ISR response which is dependent on JA, leading to the expression of JA-responsible defense-related genes, in accordance with literature data (Ahn et al., 2007; Van Loon et al., 2008; Verhagen et al., 2010; Ahn et al., 2011; Wang et al., 2013).

CONCLUSION

Recently, more and more attention of researchers has been attracted by the problem of creating bio-preparations based on beneficial strains of endophytic and rhizobacteria (Kloepper et al., 2004; Rosenblueth and Martines-Romero 2006; Bakker et. al. 2007; Yi et al., 2013). Despite the wide range of microbial preparations until now there have not been invested effective bio-fungicides for protection of crops against such harmful diseases like leaf rust, powdery mildew, leaf and glume blotch. From our point of view, one of the main causes preventing development of such preparations based on PGPRs is in the absence of systemic and integrated research of molecular mechanisms of interaction in the systems of plant-bacterial endophyte and plant-endophyte-phytopathogen. And very little is known about physiological and biochemical reactions of plants taking place during development of interaction with endophytic bacteria as well as about the mechanisms of their regulation, including participation of phytohormones (Spoel and Dong, 2008). Study of induction of plant defense responses under the influence of PGPRs may become an important component of subsequent substantiation of their effective application in agricultural practice. PGPRs are not only effective regulators of plant growth, but they are also active immuno-modulators capable of prolonged plant sensibilization and preparing their defense system for subsequent response reactions to pathogen infection. Modulating activity of all PGPRs is associated with their active production of hormone-like substances, their own antibiotics as well as elicitors (Van Loon 2007; Maksimov et al., 2011a). Alongside with different metabolites, endophytes are capable of excreting outside the cells such specific signal compounds as phytohormones, SA and JA (Dodd et.al. 2010), being triggers inducing PGPR-mediated ISR.

Establishment of PGPR-mediated ISR depends on the hormones JA and ET and requires transcriptional (co)- activators NPR1 and MYC2 (Pozo et al., 2008; Van der Ent et al., 2009; Berendsen et al., 2012). Transcription factor MYC2 (originally called JIN1 for JASMONATE INSENSITIVE1) plays a central role in JA and abscisic acid-regulated signaling (Lorenzo and Solano 2005). In the JA signaling pathway, MYC2 functions as a positive regulator of JA-responsive genes and transcription factor genes, thereby affecting the expression of a large number of JA-responsive genes. MYC2-impaired *jin1* mutants were unable to mount bacteria-mediated ISR against *P. syringae* and *H. arabidopsidis* (Pozo et al., 2008) or against *Golovinomyces orontii* (Stein et al., 2008), from which it follows that MYC2 is important regulator in priming during ISR. (Van der Ent et al., 2009). It is considered that TFs not only play the key role in transduction of signals recieved from pathogens, but fulfil cross-talk between signal pathways by either activationg or inhibiting expression of controlled genes (Kovtun et al., 2000; Mou et al., 2003; Torres 2010; Sewelam et al., 2013). JA or methyl-jasmonate can be involved in activation of the whole system of defense plant reactions important for development of resistance to abiotic and biotic stressors. Nevertheless, it remains unclear, if there exist specific reactions stimulated by jasmonate, or jasmonate itself and its derivatives bring about their effects through an increase in the content of a range of other signal intermediates, including ROS, calcium and nitrogen oxide, first of all. It remains to hope that these questions, which are important for understanding the mechanism of plant adaptation, will be clarified in the course of subsequent researches.

ACKNOWLEDGMENTS

This study was partially supported by the Ministry of Education and Science of the Russian Federation and the Russian Foundation for Basic Research by grant no. 14-04-97079.

REFERENCES

Achard, P., Renou, J.-P., Berthome, R., Harberd, N. P. & Genschik, P. (2008). Plant DELLAs restrain growth and promote survival of adversity by reducing the levels of reactive oxygen species. *Curr. Biol.,* 18, 656–660.

Ahn, I. P., Lee, S. W., Kim, M. G., Park, S. R., Hwang, D. J. & Bae, S. C. (2011). Priming by rhizobacterium protects tomato plants from biotrophic and necrotrophic pathogen infections through multiple defense mechanisms. *Mol. Cells*., 32, 7-14.

Ahn, I.-P., Lee, S.-W. & Suh, S.-C., (2007). Rhizobacteria-induced priming in Arabidopsis is dependent on ethylene, jasmonic acid, and NPR1. *Mol. Plant–Microbe Interact*., 20, 759–768.

Ali, M. B., Yu, K. W., Hahn, E. J. & Paek, K. Y. (2006). Methyl jasmonate and salicylic acid elicitation induces ginsenosides accumulation, enzymatic and non-enzymatic antioxidant in suspension culture *Panax ginseng* roots in bioreactors. *Plant Cell Reports* 25, 613–620.

Almagro, L., Gómez Ros, L. V., Belchi-Navarro, S., Bru, R., Ros Barceló, A. & Pedreño, M. A. (2009). Class III peroxidases in plant defence reactions. *J. Exp. Bot*., 60, 377-90.

Alon, M., Malka, O., Eakteiman, G., Elbaz, M., Zvi, M. M. B., Vainstein, A. & Morin, S. (2013). Activation of the Phenylpropanoid Pathway in *Nicotiana tabacum* Improves the Performance of the Whitefly Bemisia tabaci via Reduced Jasmonate Signaling. *PLoS ONE* 8(10): e76619. doi:10.1371/journal.pone.0076619.

Anderson, J. P., Badruzsaufari, E., Schenk, P. M., Manners, J. M. & Desmond, O. J. (2004). Antagonistic interaction between abscisic acid and jasmonate-ethylene signaling pathways modulates defense gene expression and disease resistance in *Arabidopsis*. *Plant Cell,* 16, 3460–79.

Anith, K. N., Momol, M. T., Kloepper, J. W., Marios, J. J., Olson, S. M. & Jones, J. B. (2004). Efficacy of plant growth-promoting rhizobacteria, acibenzolar-s-methyl and soil amendment for integrated management of bacterial wilt on tomato. *Plant. Dis.,* 88, 669–673.

Bakker, P. A. H. M , Pieterse, C. M. J. & Van Loon, L. C. (2007). Induced systemic resistance by fluorescent *Pseudomonas* spp. *Phytopathology*, 97, 239–243.

Ballare, C. L. (2011). Jasmonate-induced defenses: a tale of intelligence, collaborators and rascals. *Trends Plant Sci.,* 16, 249–57.

Bargabus, R. L., Zidack, N. K., Sherwood, J. W. & Jacobsen, B .J. (2002). Characterization of systemic resistance in sugar beet elicited by a nonpathogenic, phyllosphere-colonizing Bacillus mycoides, biological control agent. *Physiol. Mol. Plant Pathol.*, 61, 289-298.

Bari, R. & Jones, J. D. G. (2009). Role of plant hormones in plant defence responses. *Plant Mol. Biol.*, 69, 473–88.

Berendsen, R. L., Pieterse, C. M. J. & Bakker, P. A. (2012). The rhizosphere microbiome and plant health. *Trends in Plant Science*, 17, 478-486.

Bhuiyan, N. H., Selvaraj, G., Wei, Y. & King, J. (2009). Gene expression profiling and silencing reveal that monolignol biosynthesis plays a critical role in penetration defence in wheat against powdery mildew invasion. J. *Exp.Bot.,* 60, 509–521.

Burkhanova, G. F., Yarullina, L. G. & Maksivov, I. V. (2007). The control of wheat Defense responses during infection with Bipolaris sorokiniana by chitooligosaccharides. *Russian Journal of plant physiology*, 54, 119-126.

Cano-Delgado, A., Penfield, S., Smith, C., Catley, M. & Bevan, M. (2003). Reduced cellulose synthesis invokes lignification and defense responses in *Arabidopsis thaliana. Plant J.,* 34, 351–362.

Cao, S F., Zheng, Y. H., Tang, S. S. & Wang, K. T. (2008). Improved control of anthracnose rot in loquat fruit by a combination treatment of *Pichia membranifaciens* with $CaCl_2$. *International Journal of Food Microbiology*, 126, 216–220.

Chen, C., Belanger, R. R., Benhamou, N. & Paulitz, T. C. (2000). Defense enzymes induced in cucumber roots by treatment with plant growth-promoting rhizobacteria (PGPR) and *Pythium aphanidermatum. Physiological and Molecular Plant Pathology*, 56, 13–23.

Chen, L., Zhang, Z. Y., Liang, H. X., Liu, H. X., Du, L. P., Xu, H.,& Xin, Z. (2008). Overexpression of TiERF1 enhances resistance to sharp eyespot in transgenic wheat. *J. Exp. Bot.*, 59, 4195-4204.

Chet, I. & Inbar, J. (1994). Biological control of fungal pathogens. *Applied Biochem. Biotech.*, 48,. 37 – 43.

Chini, A., Fonseca, S., Fernandez, G. et al., (2007). The JAZ family of repressors is the missing link in jasmonate signalling. *Nature*, 448, 666–71.

Cohen, Y., Gisi, U. & Niederman, T. (1993). Local and systemic protection against *Phytophthora infestance* induced in potato and tomato plants by jasmonic acid and jasmonic-methyl-ester. *Phytopathology*, 83, 1054-1062.

Conrath, U., Beckers, G .J. M., Flors, V., García-Agustín, P., Jakab, G., Mauch, F., Newman, M.-A., Pieterse, C. M. J., Poinssot, B., Pozo, M. J., Pugin, A., Schaffrath, U., Ton, J., Wendehenne, W., Zimmerli, L. &

Mauch-Mani, B. (2006). Priming: getting ready for battle. *Mol. Plant–Microbe Interact.*, 19, 1062–1071.

Cosio, C. & Dunand, C. (2009). Specific functions of individual class III peroxidase genes. *Journal of Experimental Botany*, 60, 391–408.

Deepak, S., Niranjan-Raj, S., Shailasree, S., Kini, R. K., Boland, W., Shetty, H. S. & Mithöfer A. (2007). Induction of resistanse against doeny mildew pathogen in pearl millet by a synthetic jasmonate analogon. *Physiol. Mol. Plant Pathol.,* 71, 96 – 105.

De Meyer, G., Capieau, K., Audenaert, K., Buchala, A., Metraux, J. P. & Hofte, M. (1999). Nanogram amount of salicylic acid produced by rhizobacterium *P. aeruginosa* 7NSK2 activate the systemic acquired resistance pathway in bean. *Molecular Plant Microbe Interactions*, 12, 450-/458.

Demole, E., Lederer, E. & Mercier, D. (1962). Isolement et determination de la structure du jasmonate de methyle, constituant odorant caracteristique de l'essence de jasmin. *Helv Chim Acta*, 45, 675-685.

Denness, L., McKenna, J. F., Segonzac, C., Wormit, A., Madhou, P., Bennett, M., Mansfield, J., Zipfel, C. & Hamann, T. (2011). Cell wall damage-induced lignin biosynthesis is regulated by a reactive oxygen species- and jasmonic acid-dependent process in Arabidopsis. *Plant Physiol.*, 156, 1364-1374.

De Vleesschauwer, D., Djavaheri, M., Bakker, P. A. H. M. & Hofte, M. (2008). Pseudomonas fluorescens WCS374r-induced resistance in rice against *Magnaporthe oryzae* is based on pseudobactin-mediated priming for a salicylic acid-repressible multifaceted defense response. *Plant Physiology,* 148, 1996–2012.

De Vleesschauwer, D. & Höfte, M. (2009). Rhizobacteria-induced systemic resistance: 223-281. In L. C. Van Loon, Plant Innate Immunity. London, Academic Press Ltd., Elsevier Science Ltd.

De Vleesschauwer, D., Yang, Y., Cruz, C. V. & Hofte, M. (2010). Abscisic Acid-Induced Resistance against the Brown Spot Pathogen Cochliobolus miyabeanus in Rice Involves MAP Kinase-Mediated Repression of Ethylene Signaling. *Plant Physiology*, 152, 2036–2052.

Dodd, I. C., Zinovkina, N. Y., Safronova, V. I. & Belimov, A.A. (2010). Rhizobacterial mediation of plant hormone status. *Ann. Appl. Biol.*, 157, 361–379.

Doornbos, R. F., Geraats, B. P. J., Kuramae, E. E., Van Loon, L. C. & Bakker, P. A. H. M. (2011). Effects of jasmonic acid, ethylene, and salicylic acid signaling on the rhizosphere bacterial community of *Arabidopsis thaliana*. *Mol. Plant-Microbe Interact.*, 24, 395–407.

Duijff, B. J., Gianinazzi-Pearson, V. & Lemanceau, P. (1997). Involvement of the outer membrane lipopolysaccharides in the endophytic colonization of tomato roots by biocontrol *Pseudomonas fluorescens* strain WCS417r. *New Phytologist*, 135, 325–334.

Ellis, C., Karafyllidis, I., Wasternack, C. & Turner, J. G. (2002). The Arabidopsis mutant *cev1* links cell wall signalling to jasmonate and ethylene responses. *Plant Cell,* 14, 1557–1566.

Farag, M. A., Zhang, H. & Ryu, C. M. (2013). Dynamic Chemical Communication between Plants and Bacteria through Airborne Signals: Induced Resistance by Bacterial Volatiles. *J. Chem. Ecol.*, 39, 1007–1018.

Farmer, E. E. & Ryan, C. A. (1990). Interplant communication: airborne methyl jasmonate induces synthesis of proteinase inhibitors in plant leaves. *Proc. Natl. Acad. Sci. USA,* 87, 7713–7716.

Farmer, E. E. & Ryan, C. A. (1992). Octadecanoid precursors of jasmonic acid activate the synthesis of wound-inducible proteinase inhibitors. *Plant Cell,* 4, 129–134.

Fonseca, S., Chini, A., Hamberg, M., Adie, B. & Porzel, A. (2009). (+)-7-*iso*-Jasmonoyl-L-isoleucine is the endogenous bioactive jasmonate. *Nat. Chem. Biol.,* 5, 344–50.

Gadjev, I., Vanderauwera, S., Gechev, T. S., Laloi, C. & Minkov, I. N. (2006). Transcriptomic footprints disclose specificity of reactive oxygen species signaling in *Arabidopsis*. *Plant Physiology*, 141, 436–445.

Gfeller, A., Dubugnon, L., Liechti, R. & Farmer, E. E. (2010). Jasmonate biochemical pathway. *Sci. Signal.*, 16, 3(109):cm3. doi: 10.1126/scisignal.3109cm3.

Glazebrook, J. (2005). Contrasting mechanisms of defense against biotrophic and necrotrophic pathogens. *Annu. Rev. Phytopatholy*, 43, 205–227.

Hamann, T., Bennett, M., Mansfield, J. & Somerville, C. (2009). Identification of cell-wall stress as a hexose-dependent and osmosensitive regulator of plant responses. *Plant J.,* 57, 1015–1026.

Hiraga, S., Sasaki, K., Ito, H., Ohashi, Y. & Matsui, H. (2001). A large family of class III plant peroxidases. *Plant Cell Physiol.*, 42, 462-468.

Hu, X., Li, W., Chen, Q. & Yang, Y. (2009). Early signal transduction linking the synthesis of jasmonic acid in plant. *Plant Signaling and Behavior*, 4, 696-697.

Hyakumachi, M., Nishimura, M., Arakawa, T., Asano, S., Yoshida, S., Tsushima, S. & Takahashi, H. (2013). *Bacillus thuringiensis* suppresses bacterial wilt disease caused by *Ralstonia solanacearum* with systemic induction of defense-related gene expression in tomato. *Microbes Environ.,* 28, 128-34.

Jetiyanon, K. & Kloepper, J. W. (2002). Mixtures of plant growth-promoting rhizobacteria for induction of systemic resistance against multiple plant diseases. *Biol. Control,* 24, 285–291.

Jetiyanon, K., Fowler, W. D. & Kloepper, J. W. (2003). Broad-spectrum protection against several pathogens by PGPR mixtures under field conditions inThailand. *Plant Dis.*, 87, 1390–1394.

Keren-Keiserman, A., Tanami, Z., Shoseyov, O. & Ginzberg, I. (2004). Peroxidase activity associated with suberization processes of the muskmelon (*Cucumis melo*) rind. *Physiologia Plantantarum*, 121, 141–148.

Kloepper, J. W., Ryu, C. M. & Zhang, S. A. (2004). Induced systemic resistance and promotion of plant growth by *Bacillus* spp. *Phytopathology,* 94, 1259–1266.

Knoester, M., Pieterse, C. M. J., Bol, J. F. & Van Loon, L. C. (1999). Systemic resistance in *Arabidopsis* induced by rhizobacteria requires ethylene-dependent signaling at the site of application. *Mol. Plant-Microbe* Interact., 12, 720–27.

Koornneef, A. 1., Leon-Reyes, A., Ritsema, T., Verhage, A., Den Otte,r F. C., Van Loon, L. C. & Pieterse, C. M. (2008). Kinetics of salicylate-mediated suppression of jasmonate signaling reveal a role for redox modulation. *Plant Physiol.*, 147, 1358-1368.

Kovtun, Y., Chiu, W. L., Tena, G. & Sheen, J. (2000). Functional analysis of oxidative stress-activated mitogen-activated protein kinase cascade in plants. *Proc. Natl. Acad. Sci. USA*, 97, 2940–2945.

Kumari, G. J., Reddy, A. M., Naik, S. T., Kumar, S. G., Prasanthi, J., Sriranganayakulu, G., Reddy, P. C. & Sudhakar, C. (2006). Jasmonic acid induced changes in protein pattern, antioxidative enzyme activities and peroxidase isoenzymes in peanut seedlings. *Biologia Plantarum*, 50, 219–226.

Kwak, J. M., Mori, I. C., Pei, Z. M., Leonhardt, N., Torres, M. A., Dangl, J. L., Bloom, R. E., Bodde, S., Jones, J. D. & Schroede,r J. I. (2003). NADPH oxidase AtrbohD and AtrbohF genes function in ROS-dependent ABA signaling in Arabidopsis. *Embo J.,* 22, 2623–2633,

Li, J., Zingen-Sell, I. & Buchenauer, H. (1996). Iduction of resistance of cotton plants to Verticillium wilt and of tomato plants to Fusarium wilt by

3-aminobutyric acid and methyl jasmonate. *J. Plant Dis. Prot.*, 103, 288–299.

Lin, W., Anuratha, C. S., Datta K. et al., (1995). Genetic engineering of rice for resistance to sheath blight. *Bio. Technology*, 13, 686–691.

Lorenzo, O., Chico, J. M., Sanchez-Serrano, J. J. & Solano, R. (2004). *JASMONATE-INSENSITIVE1* encodes a MYC transcription factor essential to discriminate between different jasmonate-regulated defense responses in *Arabidopsis*. *Plant Cell*, 16, 1938–1950.

Lorenzo, O., Piqueras, R., Sanchez-Serrano, J. J. & Solano, R. (2003). ETHYLENE RESPONSE FACTOR1 integrates signals from ethylene and jasmonate pathways in plant defense. *Plant Cell*, 15, 165–178.

Lorenzo, O. & Solano, R. (2005). Molecular players regulating the jasmonate signaling network. *Curr. Opin. Plant Biol.*, 8, 532–540.

Maksimov, I. V., Abizgildina, R. R. & Pusenkova, L. I. (2011a). Plant growth Promoting Microorganisms as Alternanive to Chemical Protection from Pathogens. *Russian Journal Appl. Biochem. Microbiol.*, 47, 333–345.

Maksimov, I. V., Abizgildina, P. P., Sorokan, A. V. & Burkhanova, G. F. (2014). Regulation of Peroxidase Activity under the Influence of SignalingMolecules and *Bacillus subtilis* 26D in Potato Plants Infectedwith *Phytophthora infestans*. *Russian Journal Applied Biochemistry and Microbiology*, 50, 173–178.

Maksimov, I. V., Cherepanova, E. A., Burkhanova, G. F., Sorokan, A. V. & Kuzmina, O. I. (2011b). Structural-functional features of plant isoperoxidases. *Biochemistry-Moscow*, 76, 749-763.

Maksimov, I. V., Cherepanova, E. A., Yarullina, L. G. & Akhmetova, I. E. (2005). Isolation of chitin-specific oxidoreductases. *Russian Journal Appled Biochemistry and Microbiology*, 41, 616-620.

Mandal, S., Kar, I., Mukherjee, A. K. & Acharya, P. (2013). Elicitor-induced defense responses in *Solanum lycopersicum* against *Ralstonia solanacearum*. *Scientific World Journal*, 25, 561056. doi: 10.1155/2013/561056. eCollection 2013.

Meyer, A., Miersch, O., Buettner, C., Dathe, W. & Sembdner, G. (1984). Occurrence of the plant growth regulator jasmonic acid in plants. *J. Plant Growth Regul.*, 3, 1-8.

Memelink, J. (2009). Regulation of gene expression by jasmonate hormones. *Phytochemistry*, 70, 1560–1570.

Miedes, E., Vanholme, R., Boerjan, W. & Molina, A. (2014). The role of the secondary cell wall in plant resistance to pathogens. *Front Plant Sci.*, 5, 5:358. doi: 10.3389/fpls.2014.00358. eCollection 2014.

Mika, A., Boenisch, M. J., Hopff, D. & Luthje, S. (2010). Membrane-bound guaiacol peroxidases from maiz (*Zea mays* L.) roots are regulated by methyl jasmonate, salicylic acid, and pathogen elicitors. *Journal of Experimental Botany,* 61, 831 – 841.

Mou, Z., Fan, W. & Dong, X. (2003). Inducers of plant systemic acquired resistance regulate NPR1 function through redox changes. *Cell,* 113, 935–944.

Nakkeeran, S., Kavitha, K., Chandrasekar, G., Renukadevi, P. & Fernando W. G. D. (2006). Induction of plant defence compounds by *Pseudomonas chlororaphis* PA23 and *Bacillus subtilis* BSCBE4 in controlling damping-off of hot pepper caused by *Pythium aphanidermatum*. *Biocontrol Science and Technology*, 16, 403-/416.

Okazaki, Y., Ishizuka, A., Ishihara, A., Nishioka, T. & Iwamura, H. (2007). New dimeric compounds of avenanthramide phytoalexin in oats. *Journal of Organic Chemistry,* 72, 3830–3839.

Oliver, R. P. & Ipcho, S. V. (2004). Arabidopsis pathology breathes new life into the necrotrophs-vs.-biotrophs classification of fungal pathogens. *Mol Plant Pathol.*, 1, 347-352.

Orozco-Cardenas, M. L., Narvaez-Vasquez, J. & Ryan, C. A. (2001). Hydrogen peroxide acts as a second messenger for the induction of defense genes in tomato plants in response to wounding, systemin, and methyl jasmonate. *Plant Cell,* 13, 179–191.

Overmyer, K., Brosche, M. & Kangasjoarvi, J. (2003). Reactive oxygen species and hormonal control of cell death. *Trends Plant Sci.,* 8, 335–342.

Passardi, F., Cosio, C., Penel, C. & Dunand, C. (2005). Peroxidases have more functions than a Swiss army knife. *Plant Cell Reports*, 24, 255–265.

Pieterse, C. M. J., Leon-Reyes, A., Van der Ent, S. & Van Wees, S. C. M. (2009). Networking by small-molecules hormones in plant immunity. *Nat. Chem. Biol.*, 5, 308–316.

Pieterse, C. M. J., Van der Does, D., Zamioudis, C., Leon-Reyes, A. & Van Wees, S. C. M. (2012). Hormonal Modulation of Plant Immunity. *Annu. Rev. Cell Dev. Biol.,* 28, 28.1–28.33.

Pieterse, C. M. J., Van Pelt, J. A., Ton, J., Parchmann, S., Mueller, M. J., Buchala, A. J., Métraux, J.-P. & Van Loon, L. C. (2000). Rhizobacteria-mediated induced systemic resistance (ISR) in Arabidopsis requires sensitivity to jasmonate and ethylene but is not accompanied by an increase in their production. *Physiol. Mol. Plant Pathol.*, 57, 123–134.

Pieterse, C. M. J., Van Wees, S. C. M., Van Pelt, J. A., Knoester, M., Laan, R., Gerrits, H., Weisbeek, P. J. & Van Loon, L. C., (1998). A novel signaling pathway controlling induced systemic resistance in Arabidopsis. *Plant Cell,* 10, 1571–1580.

Podile, A. R. & Laxmi, V. D. V. (1998). Seed bacterization with *B. subtilis* AFI increase PAL and reduces the incidence of fusarial wilt in pigeon pea. *Journal of Phytopathology*, 146, 255-/259.

Pozo, M. J., Van der Ent, S., Van Loon, L. C. & Pieterse, C. M. J. (2008). Transcription factor MYC2 is involved in priming for enhanced defense during rhizobacteria-induced systemic resistance in *Arabidopsis thaliana*. *New Phytol.*, 180, 511–523.

Quan, L.-J., Zhang, B., Shi, W.-W. & Li, H.-Y. (2008). Hydrogen peroxide in plants: a versatile molecule of the reactive oxygen species network . J. Integ. *Plant Biology*, 50, 2 – 18.

Raupach, G. S. & Kloepper, J. W. (1998). Mixtures of plant growth-promoting rhizobacteria enhance biological control of multiple cucumber pathogens. *Phytopathology,* 88, 1158–1164.

Repka, V., Fischerova, I. & Silharova, K. (2004). Methyl jasmonate is a potent elicitor of multiple defence responses in grapevine leaves and cell-suspension cultures. *Biologia Plantarum*, 48, 273–283.

Robert-Seilaniantz, A., Grant, M. & Jones, J. D. G. (2011). Hormone crosstalk in plant disease and defense: more than just jasmonate-salicylate antagonism. *Annu. Rev. Phytopathol.*, 49, 317–343.

Romero, D., Rivera, M. E., Cazorla, F. M., Codina, J. C., Fernández-Ortuño, D., Torés, J. A., Pérez-García, A., & de Vicente A. (2008). Comparative histochemical analyses of oxidative burst and cell wall reinforcement in compatible and incompatible melon-powdery mildew (*Podosphaera fusca*) interactions. *J. Plant Physiol.*, 165, 1895–1905.

Ros Barcelo, A., Gomez Ros, L. V. & Esteban Carrasco, A. (2007). Looking for syringyl peroxidases. *Trends in Plant Scienc*, 12, 486–491.

Rosenblueth, M. & Martínez-Romero, E. (2006). Bacterial Endophytes and Their Interactions with Hosts. *Molecular Plant–Microbe Interactions*, 19, 827–837.

Ryals, J. A., Neuenschwander, U. H., Willits, M. G., Molina, A., Steiner, H. Y. & Hunt, M. D. (1996). Systemic acquired resistance. *Plant Cell*, 8, 1809-1819.

Salim, A., Saminaidu, K., Marimuthu, M., Perumal, Y., Rethinasamy, V. & Palanisami, J. (2011). Defense responses in tomato landrace and wild genotypes to early blight pathogen *Alternaria solani* infection and accumulation of pathogenesis-related proteins. *Arch. Phytopathol. Plant.* Prot., 44, 1147-1164.

Sasaki, K., Iwai, T., Hiraga, S., Kuroda, K., Seo, S., Mitsuhara, I., Miyasaka, A., Iwano, M., Ito, H., Matsui, H. & Ohashi, Y. (2004). Ten rice peroxidases redundantly respond to multiple stresses including infection with rice blast fungus. *Plant Cell Physiol.*, 45, 1442-1452.

Segarra, G., Van der Ent, S., Trillas, I. & Pieterse, C. M. J. (2009). MYB72, a node of convergence in induced systemic resistance triggered by a fungal and a bacterial beneficial microbe. *Plant Biology*, 11, 90–96.

Seo, S., Seto, H., Yamakawa, H. & Ohashi, Y. (2001).Transient accumulation of jasmonic acid during the synchronized hypersensitive cell death in *Tobacco mosaic virus*-infected tobacco leaves. *Mol. Plant-Microbe Interact.,* 14, 261-264.

Sewelam, N., Kazan, K., Thomas-Hall, S. R., Kidd, B. N., Manners, J. M. & Schenk, P. M. (2013). Ethylene Response Factor 6 Is a Regulator of Reactive Oxygen Species Signaling in *Arabidopsis*. *PLoS ONE* 8(8): e70289. doi:10.1371/journal.pone.0070289

Shadle, G. L., Wesley, S. W., Korth, K. L., Chen, F., Lamb, C., & Dixon, R. A. (2003). Phenylpropanoid compounds and disease resistance in transgenic tobacco with altered expression of L-phenylalanine ammonia-lyase. *Phytochemistry,* 64, 153–161.

Spoel, S. H. & Dong, X. (2008). Making sense of hormone crosstalk during plant immune responses. *Cell Host Microbe*, 3, 348–351.

Staswick, P. E. & Tiryaki, I. (2004). The oxylipin signal jasmonic acid is activated by an enzyme that conjugates it to isoleucine in *Arabidopsis*. Plant Cell, 16, 2117–2127.

Stein, E., Molitor, A., Kogel, K. H. & Waller, F. (2008). Systemic resistance in *Arabidopsis* conferred by the mycorrhizal fungus *Piriformospora indica* requires jasmonic acid signaling and the cytoplasmic function of NPR1. *Plant Cell Physiol.*, 49, 747–751.

Stepanova, A. N. & Ecker, J. R. (2000). Ethylene signaling: from mutants to molecules. *Curr. Opin. Plant Biol.*, 3, 353–360.

Sticher, L., Mauch-Mani, B. & Métraux, J. P. (1997). Systemic acquired resistance. *Annu. Rev. Phytopathol.*, 35, 235-270.

Tabei, Y., Kitade, S., Nishizawa, Y., Kikuchi, N., Kayano, T., Hibi, T. & Akutsuet, K. (1998). Transgenic cucumber plants harboring a rice chitinase gene exhibit enhanced resistance to gray mold (*Botrytis cinerea*). *Plant Cell Reports*., 17, 159–164.

Thomma, B. P. H. J., Eggermont, K., Penninckx, I. A., Mauch-Mani, B., Vogelsang, R., Cammue, B. P. A. & Willem F. (1998). Broekaert Separate jasmonate-dependent and salicylate-dependent defense-response pathways in Arabidopsis are essential for resistance to distinct microbial pathogens. *Proc. Natl. Acad. Sci. USA*, 95, 15107–15111.

Torres, M. A. (2010). ROS in biotic interactions. *Physiol Plant.*, 138, 414-429.

Traw, M. B., Kim, J., Enright, S., Cipollini, D. F. & Bergelson J. (2003). Negative cross-talk between salicylate- and jasmonate- mediated pathways in the Wassilewskija ecotype of *Arabidopsis thaliana*. *Molecular Ecology,* 12, 1125 – 1135.

Ueda, J. & Kato, J. (1980). Isolation and Identification of a Senescence-promoting Substance from Wormwood (*Artemisia absinthium* L.). *Plant Physiol.,* 66, 246-249.

Van der Ent, S., Verhagen, B. W. M.,Van Doorn, R., Bakker, D., Verlaan, M.G., Pel, M. J. C., Joosten, R. G., Proveniers, M. C.G., Van Loon, L.C., Ton, J. & Pieterse C. M. J. (2008). MYB72 is required in early signaling steps of rhizobacteria-induced systemic resistance in *Arabidopsis*. *Plant Physiol.*, 146, 1293–1304.

Van der Ent, S., Van Wees, S. C. M. & Pieterse, C. M. J. (2009). Jasmonate signaling in plant interactions with resistance-inducing beneficial microbes. *Phytochemistry,* 70, 1581–1588.

Vanholme, R., Demedts, B., Morreel, K., Ralph, J. & Boerjan, W. (2010). Lignin biosynthes is and structure. *PlantPhysiol.*, 153, 895–905.

Van Loon, L. C. (2007). Plant responses to plant growth-promoting rhizobacteria. *Eur. J. Plant Pathol.*, 119, 243-254.

Van Loon, L. C., Bakker, P. A. H. M. & Pieterse, C. M. J. (1998). Systemic resistance induced by rhizosphere bacteria. *Annual Review of Phytopathology,* 36, 453–483.

Van Loon, L. C., Bakker, P. A. H. M., Van Der Heijdt, W. H. W., Wendehenne, D. & Pugin, A. (2008). Early responses of tobacco suspension cells to rhizobacterial elicitors of induced systemic resistance. *Molecular Plant–Microbe Interactions*, 21, 1609–1621.

Van Loon, L. C., Rep, M. & Pieterse, C. M. J. (2006). Significance of inducible defense-related proteins in infected plants. *Annu. Rev. Phytopathol.,* 44, 135–162.

Van Oosten, V. R., Bodenhausen, N., Reymond, P., Van Pelt, J. A., Van Loon, L. C., Dicke, M. & Pieterse, C. M. J. (2008). Differential effectiveness of microbially induced resistance against herbivorous insects in Arabidopsis. *Mol Plant Microbe Interact.*, 21, 919–930.

Van Peer, R., Niemann, G. J. & Schippers, B. (1991). Induced resistance and phytoalexin accumulation in biological control of *Fusarium* wilt of carnation by *Pseudomonas* sp. WCS417r. *Phytopathology,* 81, 728–734.

Van Wees, S. C. M., De Swart, E. A. M., Van Pelt, J. A., Van Loon, L. C. & Pieterse, C. M. J. (2000). Enhancement of induced disease resistance by simultaneous activation of salicylate- and jasmonate-dependent defense pathways in *Arabidopsis thaliana. Proc. Natl. Acad. Sci. USA,* 97, 8711–8716.

Van Wees, S. C. M., Luijendijk, M., Smoorenburg, I., Van Loon, L. C. & Pieterse, C. M. J. (1999). Rhizobacteria-mediated induced systemic resistance (ISR) in *Arabidopsis* is not associated with a direct effect on expression of known defense-related genes but stimulates the expression of the jasmonate-inducible gene *Atvsp* upon challenge. *Plant Mol. Biol.,* 41, 537–49.

Van Wees, S. C. M., Van der Ent, S. & Pieterse, C. M. J. (2008). Plant immune responses triggered by beneficial microbes. *Curr. Opin. Plant. Biol.,* 11, 443–448.

Verhagen, B. W. M., Glazebrook, .J, Zhu, T., Van Loon, L. C. & Pieterse, C. M. J. (2004). The transcriptome of rhizobacteria-induced systemic resistance in Arabidopsis. *Molecular Plant–Microbe Interactions*, 17, 895–908.

Verhagen, B W. M., Trotel-Aziz1, P., Couderchet, M., Hofte, M. & Aziz, A. (2010). *Pseudomonas* spp.-induced systemic resistance to *Botrytis cinerea* is associated with induction and priming of defence responses in grapevine. *Journal of Experimental Botany*, 61, 249–260.

Vijayan, P., Shockey, J., Lévesque, C. A., Cook, R. J. & Browse, J. (1998). A role for jasmonate in pathogen defense of *Arabidopsis. Proc. Natl. Acad. Sci. USA,* 95, 7209–7214.

Vos, I. A., Pieterse, C. M. J. & Van Wees, S. C. M. (2013). Costs and benefits of hormone-regulated plant defences. *Plant Pathology*, 62, 43–55.

Wang, X., Wang, J., Jin, P. & Zheng, Y. (2013). Investigating the efficacy of *Bacillus subtilis* SM21 on controlling *Rhizopus* rot in peach fruit. *International Journal of Food Microbiology*, 164, 141–147.

Way, H. M., Birch, R. G. & Manners, J. M. (2011). A comparison of individual and combined L-phenylalanine ammonia-lyase and cationic peroxidase transgenes for engineering resistance in tobacco to necrotrophic pathogens. *Plant Biotechnol. Rep.*, 5, 301–308.

Way, H. M., Kazan, K., Mitter, N., Goulter, K. C., Birch, R. G. & Manners, J. M. (2002). Constitutive expression of a phenylalanine ammonia-lyase gene from *Stylosanthes humilis* in transgenic tobacco leads to enhanced disease resistance but impaired plant growth. *Physiol. Mol. Plant Pathol.,* 60, 275–282.

Wei, G., Kloepper, J. W. & Tuzun, S. (1991). Induction of systemic resistance of cucumber to *Colletotrichum orbiculare* by select strains of plant growthpromoting rhizobacteria. *Phytopathology,* 81, 1508–1512.

Wu, C.T. & Bradford, K.J. (2003). Class I chitinase and β-1.3-glucanase are differentially regulated by wounding, methyl jasmonate, ethylene, and gibberellin in tomato seeds and leaves. *Plant Physiol.*, 133, 263-73.

Xu, L., Zhu, L., Tu, L., Liu, L., Yuan, D., Jin, L., Long, L. & Zhang X. (2011). Lignin metabolism has a central role in the resistance of cotton to the wilt fungus *Verticillium dahliae* as revealed by RNA-Seq-dependent transcription alanalysis and histochemistry. *J. Exp.Bot.*, 62, 5607–5621.

Yi, H. S., Yang, J. W. & Ryu, C. M. (2013). ISR meets SAR outside: additive action of the endophyte *Bacillus pumilus* INR7 and the chemical inducer, benzothiadiazole, on induced resistance against bacterial spot in field-grown pepper. *Front. Plant Sci.*, 4:122. doi: 10.3389/fpls.2013.00122.

In: Jasmonic Acid
Editor: Lucille Morrison
ISBN: 978-1-63463-854-8

Chapter 4

JASMONIC ACID: ROLE IN PLANT SYSTEMIC ACQUIRED RESISTANCE (SAR)

Chithrashree and C. Srinivas*
Biocontrol Research Laboratory,
Department of Microbiology and Biotechnology,
Bangalore University, Jnana Bharathi Campus,
Bangalore, Karnataka, India

ABSTRACT

Plant growth regulators are not only involved in regulating developmental processes in plants but also play important roles for the plant's responses to biotic and abiotic stresses. Jasmonic acid (JA) is a naturally occurring plant growth regulator biosynthesized from linolenic acid by the octadecanoid pathway. The major function of JA is to regulate plant growth responses to abiotic and biotic stresses as well as plant growth and development. Systemic acquired resistance (SAR) is a potent innate immunity system in plants that is effective against a broad range of pathogens. Accumulation of Jasmonic acid (JA) and Salicylic acid (SA) in non-infected part renders the whole plant more resistant to subsequent pathogen infection. Mobile signals for systemic acquired resistance are translocated via vasculature and may include small peptides such as

* Corresponding author: Dr. C. Srinivas, Department of Microbiology and Biotechnology, Bangalore University, Jnana Bharathi Campus, Bangalore 560 056, Karnataka, India. Tel: +91-080-22961461, Fax: +91-080-23219295.

methyl salicylate, jasmonate, azelaic acid and other lipid-derived molecules. The JA signal pathway involves several signal transduction events: the perception of the infection or stress stimulus and transduction of the signal locally and systemically; the perception of this signal and induction of JA biosynthesis; the perception of JA and induction of responses; and finally, integration of JA signaling with outputs from the SA, ethylene, and other signaling pathways. Although the JA, SA and ethylene signaling pathways are clearly defined by the signaling molecules they synthesize, they interact cooperatively and antagonistically in a variety of responses. This chapter focuses on biosynthesis of Jasmonic acid and its role in whole plant resistance or systemic induced resistance.

Introduction

Jasmonates are a growing class of signaling molecules and plant hormones which are derived from polyunsaturated fatty acids via the octadecanoid pathway, and characterized by a pentacyclic ring structure. Like other plant hormones, jasmonates exhibit a broad spectrum of physiological activities, ranging from seed germination over reproductive development, all the way to senescence. Jasmonates also serve important roles as signaling molecules in plant defense, particularly defense against insect herbivore and necrotrophic pathogens.

Since the identification of methyl jasmonate (MeJA) as a secondary metabolite in essential oils of jasmine in 1962 (Demole et al. 1962), many related compounds have been discovered in a wide range of plants, and are collectively referred to as jasmonates. Jasmonates share their biosynthetic origin from oxygenated polyunsaturated fatty acids. In the early 1980s, their widespread occurrence throughout the plant kingdom (Meyer et al. 1984), and their growth-inhibitory (Dathe et al. 1981), and senescence-promoting activities (Ueda and Kato 1980) were established. First indications for a role of jasmonates in the regulation of gene expression were obtained by Parthier and co-workers who observed the accumulation of jasmonate inducible proteins (JIPs) in senescing barley leaves (Weidhase et al., 1987; Mueller- Uri et al. 1988). Shortly after, jasmonates were shown to mediate the elicitor-induced accumulation of antimicrobial phytoalexins in cell cultures (Gundlach et al., 1992), and the induction of vegetative storage proteins in soybean and *Arabidopsis* (Franceschi and Grimes, 1991; Staswick et al., 1992). These findings greatly stimulated the interest in jasmonates as a new class of signaling molecules in plant defense against both insects and pathogens.

In recent years, numerous mutants were characterized that are impaired in either jasmonate synthesis or response clearly establishing their function as plant-defense regulators. Mutants impaired in jasmonate perception and signaling, including the *coi1*, *jin1* and *jai3* mutants in Arabidopsis (Lorenzo et al., 2004; Lorenzo and Solano, 2005) and *jai1* mutant in tomato (Lie et al., 2004) fail to mount appropriate defense responses. The mutants that are deficient in the polyunsaturated fatty acid precursors of jasmonates also comprised in the induction of defense responses (McConn et al., 1997; Lie et al., 2003).

Jasmonic Acid

Jasmonic acid and its various metabolites are involved in plant cell signaling responses to abiotic and biotic stress as well as plant growth and development. JA is involved in the regulation of processes that include growth inhibition, senescence, flower development, leaf abscission and tendril coiling. At concentrations >50 μM, JA induces senescence in plant cell cultures and excised leaves. The senescence response includes a loss of chlorophyll, degradation of chloroplast proteins such as ribulose bisphosphate carboxylase, and accumulation of new proteins. Tendril coiling is induced by mechanical stimulation and roles for auxin and ethylene have been proposed. Jasmonate has been shown to stimulate fruit ripening, most likely through its action on ethylene biosynthesis (Czapski and Saniewski, 1992). In addition, jasmonate stimulates the conversion of lycopene to 3-carotene (Saniewski and Czapski, 1983).

Since discovered in the 1960s as secondary metabolites from the oils of jasmine flowers (Demole et al., 1962), the biological roles of JA have received increased attention of researchers in the past decades. Jasmonates have gradually become realized as a defense and fertility hormone, and as such modulate numerable processes relating to development and stress responses. In Arabidopsis and tomato, JAs are directly involved in stamen and trichome development, vegetative growth, cell cycle regulation, senescence, anthocyanin biosynthesis regulation, and responses to various biotic and abiotic stresses (Wasternack, 2007; Howe and Jander, 2008; Browse, 2009; Avanci et al., 2010; Pauwels and Goossens, 2011). In monocots, much less is known about the role of JAs, however, it has been shown they are required for sex determination, reproductive bud initiation and elongation, leaf senescence, pigmentation of tissues and responses to the attack by pathogens and insects (Engelberth et al., 2004; Tani et al., 2008; Acosta et al., 2009; Yan et al, 2012).

In plants, the JA signal acts co-operatively with other plant hormones. A number of studies have already attracted attention to plant hormone cross-talk as it relates to defense responses. In Arabidopsis, JA was shown to interact synergistically with ethylene (Xu et al., 1994), and, depending on particular stress, both synergistically and antagonistically with salicylic acid (Beckers and Spoel, 2006) and absicisic acid (ABA) (Anderson et al., 2004) in plant-pathogen or insect interactions. Gibberellins (GA) interact with JA to control flower fertility in Arabidopsis. In maize, JA positively regulates ABA and ET biosynthesis in senescing leaves (Yan et al., 2012). In review it is clear that JA signaling exert its functions through interaction with multiple plant hormones; however the junction of these interactions still remain to be explored.

In addition to its role in plant growth and development, JA has been proposed as a key regulator of plant responses to abiotic and biotic stresses including necrotrophic pathogen infection, wounding and insect attack elicit signals that trigger a phosphorylation cascade leading to jasmonic acid (JA) biosynthesis. JA is then detected by receptors which activate a signal transduction pathway involved in the expression of JA-responsive genes, resulting in the formation of chemical and physical barriers against the pathogen or herbivore (Kazan and Manners, 2008), but also reduced overall growth. Activation of JA signaling by exogenous supply of methyl jasmonate (MeJA) has also been shown to increase the release of signaling compounds such as flavonoids and indoles from plant roots (Buer et al., 2010; Faure et al., 2009). Low concentrations of JA were found to induce genes encoding proteinase inhibitors to control insects, enzymes involved in flavonoid biosynthesis (chalcone synthase and phenylalanine ammonia lyase) (Creelman et al., 1992), sesquiterpenoid biosynthesis (hydroxymethylglutaryl CoA reductase) (Choi et al., 1992), thionin (antifungal protein) (Andresen et al., 1992), osmotin (antifungal protein) and defensin (Xu et al., 1994). Jasmonate also induces expression of lipoxygenase (Bell and Mullet, 1991). Lipoxygenase is involved in the biosynthesis of jasmonate and has been implicated in plant responses to pathogens (Peever and Higgins, 1989). The induction of lipoxygenase may make the plant more sensitive to further pathogen attack or enhance the plants' capacity to synthesize other lipid-derived compounds used in plant defense. JA and ethylene induce PR-3, PR-4, PDF1.2, chitinases (CHI-B), and hevein-like protein.

JA has been found in all higher plants examined and in general is estimated to be present at concentrations <10 μM (Manson et al., 1992). The level of JA in plant tissues varies as a function of tissue type, development, and external stimuli (Sembdner and Parthier, 1993). The highest levels of JA / JAMe (methyl jasmonate) are reported in flowers and reproductive tissues, whereas much lower levels are found in roots and mature leaves. Jasmonate can move readily in plants in the liquid and vapor phase. Changes in plant gene expression are induced by nanomolar to micromolar concentrations of JA/JAMe. Numerous derivatives of JA are found in plants including esters such as methyl jasmonate and hydroxylated forms such as tuberonic acid and cucurbic acids. JA can be conjugated to amino acids, i.e. leucine, valine, isoleucine and the sugar, B-glucoside using UDP-glucose. JA can influence several aspects of plant growth and development. At concentrations >50 μM, JA induces senescence in plant cell cultures and excised leaves.

Jasmonic acid Biosynthesis

Jasmonates are derived from a family of oxygenated fatty acids collectively known as oxylipins, which can be made internally from enzymatic reactions within the plant cell (Wasternack and kombrick, 2010) or derived externally from a few fungal species (Miersch et al., 1999). Plant oxylipins can be formed enzymatically, by initial oxidation by lipoxygenases (LOXs) or α-dioxygenases (α-DOXs); however, non-enzymatic autoxidation of poly-unsaturated fatty acids (PUFA) also contribute to oxylipins formation in plant (Gobel and Feussner, 2009). An array of these substances are known to exert protective activities either as signaling molecules in plants during development, wounding, and insect and pathogen attack, or direct anti-microbial substance that are toxic to the invader.

JA biosynthesis and signaling pathways have been extensively investigated in dicotyledonous plants such as Arabidopsis, tobacco and tomato. In monocotyledonous species, only a scant number of JA biosynthetic enzymes have been described (Tani et al., 2008; Yan et al., 2012). Jasmonates are synthesized by the conversion of ά-linolenic acid to 12-oxo-phytodienoic acid (OPDA) by a series of enzymatic reactions in the chloroplast (Kazan and Manners, 2008). OPDA is then transported to the peroxisomes where it undergoes a series of β-oxidation reactions to generate JA (Wasternack, 2007; Kazan and Manners, 2008). JA can be further enzymatically converted into numerous derivatives or conjugates, some of which have well-described biological activity such as free JA, MeJA, cis-jasmone and JA–Ile (Figure 5). Biosynthesis of JAs originates from polyunsaturated fatty acids (PUFA) and is synthesized by one of the seven distinct branches of the lipoxygenase (LOX) pathway, the allene oxide synthase (AOS) branch (Feussner and Wasternack, 2002).

Conversion of PUFAs to OPDA in Chloroplast

The original precursors Polyunsaturated fatty acids (PUFAs) including linoleic acid (18:2), linolenic acid (18:3) and hexadecatrienoic acid (16:3) are released from chloroplast membranes by the action of lipid hydrolyzing enzymes and are readily oxidized to yield the corresponding fatty acid hydroperoxides.

Under conditions of oxidative stress hydroperoxides are formed by free-radical-catalyzed oxidation of PUFAs (Mueller, 2004). Alternatively, fatty acid hydroperoxides are synthesized enzymatically involving lipoxygenase (LOX) or ά-dioxygenase (DOX) activities. The LOX catalyzed regio- and stereo-specific dioxygenation of PUFAs at C9 or C13 results in the specific formation of 9(S) and 13(S) hydroperoxy fatty acids, respectively.

These hydroperoxides, 13-HPOT (13S-hydroperoxy-(9Z, 11E,15)-octadecatrienoic acid) are substrates for at least seven different families of enzymes, resulting in the formation of different classes of oxylipins including jasmonates (Figure 1) (Mosblech et al., 2009; Wasternack, 2007).

The first committed step of JA biosynthesis i.e. the octadecanoid is performed by allene oxide synthase (AOS), an unusual cytochrome P450 which does not bind molecular oxygen but uses already oxygenated fatty acid hydroperoxide substrates as oxygen donor and as source for reducing equivalents (Howe and Schilmiller 2002; Werck-Reichhart et al., 2002) (Figure 2). AOS catalyzes the dehydration of 13(S)-hydroperoxy-octadecatrienoic acid (13-HPOT) to form an unstable allene oxide 12,13(S)-epoxy-octadecatrienoic acid (12,13-EOT). In aqueous media, 12, 13-EOT rapidly decomposes to α and γ ketols, or undergoes cyclization to form 12-oxy-phytodienoic acid (OPDA). As opposed to spontaneous cyclization which results in a racemic mixture of OPDA enantiomers, allene oxide cyclase (AOC) ensures the formation of optically pure enantiomer, 9S, 13S/*cis* (+)-OPDA. The short life of 12, 13-EOT (Hamberg and Fahlstadius 1990; Ziegler et al., 1999) and the optical purity of endogenous OPDA (Laudert et al., 1997) suggest tight coupling of the AOS and AOC reactions in vivo. However, physical contact of AOS and AOC in an enzyme complex does not seem to be required for cyclization reaction (Zerbe et al., 2007). The formation of 9S, 13S-OPDA as the first member of the jasmonate family with signaling activity concludes the plastid-localized part of the octa (hexa) decanoid pathway. All the reactions from α-linolenic acid to OPDA take place within a plastid and later *cis* (+)-OPDA, which is one out of four possible OPDA stereoisomers and precursor for biologically active JA, is subsequently transported to perioxisomes, where the final steps of JA biosynthesis occur. ABC transporter COMATOSE (CTS) catalyses the ATP-dependent import of fatty acids into perioxisomes as substrates for β-oxidation (Figure 2).

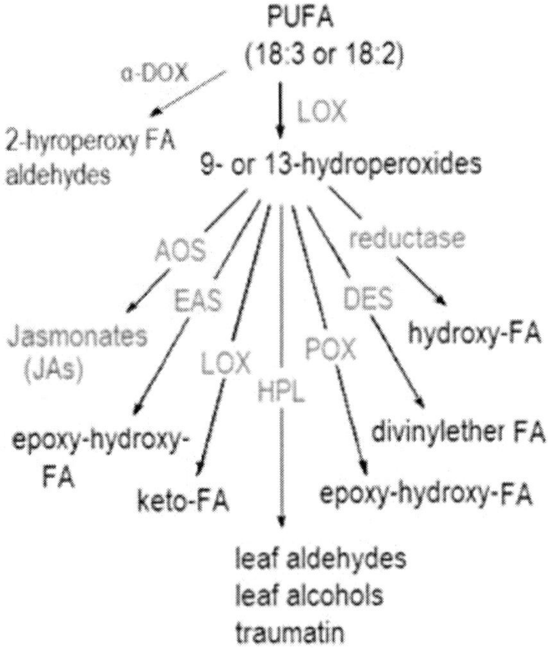

Source: Andreou et al., 2009.

Figure 1. Overview of the oxylipins biosynthesis pathways in plants. In the first reaction, free fatty acids (18:3 or 18:2) are oxidized by the addition of molecular oxygen yielding hydroperoxides (HPOT, hydroperoxy-octadecatrienoic acid; HOPD, hydroperoxyoctadecadienoic acids) through activity of oxygenases, lipoxygenase (LOX) or α-dioxygenase (α-DOX). Hydroperoxide products formed by LOXs are further metabolized by other enzymes: allene oxide synthase (AOS), hydroperoxide lyase (HPL), divinyl ether synthase (DES), peroxygenase (POX) and epoxyalcohol synthases (EAS).

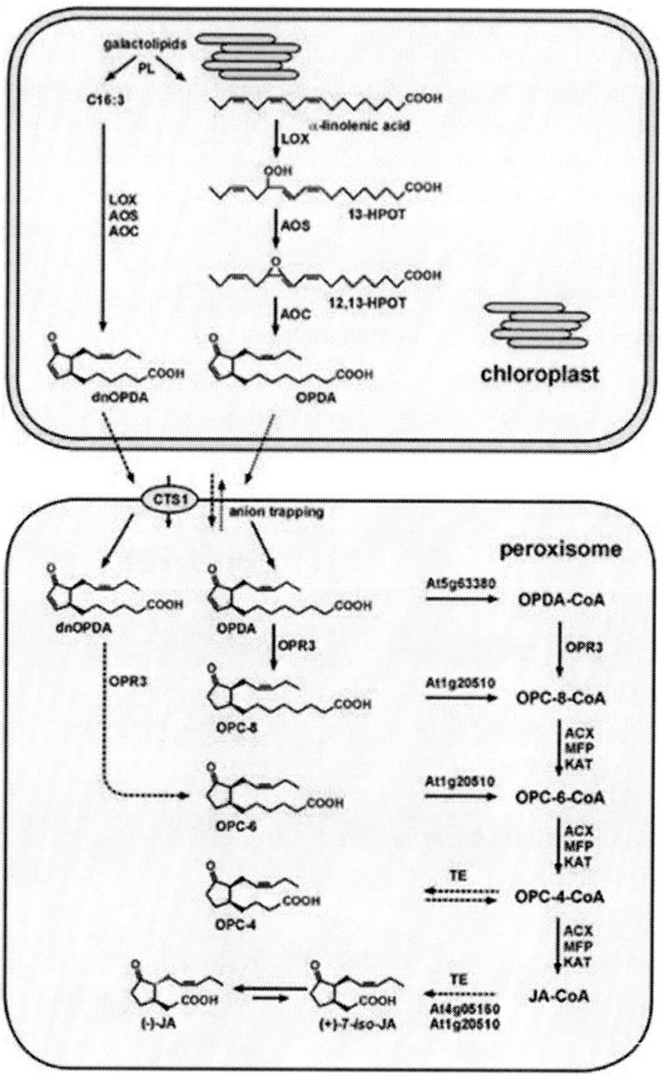

Source: Kienow et al., 2008.

Figure 2. Biosynthesis of jasmonic acid through the octadecanoid pathway. Polyunsaturated fatty acids (18:3 and 16:3) are precursors for jasmonic acid biosynthesis. The first three steps are localized in plastids and lead to the formation of 12-oxophytodienoic acid (OPDA) and dinor 12-oxophytodienoic acid (dnOPDA) by the consecutive action of plastid-localized lipoxygenase (LOX), allene oxide synthase (AOS), and allene oxide cyclase (AOC). The subsequent steps, including the reduction of OPDA (dnOPDA) to OPC8:0 (OPC6:0) followed by three (two) cycles of β-oxidation result in the production of jasmonic acid in peroxisomes.

The cts mutants with reduced levels of JA, is impaired in wound-induced JA accumulation, and expresses the JA-dependentVSP1 gene at a lower level suggesting that CTS delivers substrates for JA synthesis into the peroxisomes, most likely (dn)OPDA, or the corresponding CoA esters (Theodoulou et al., 2005). Additional pathways for fatty acid import into perioxisomes, by diffusion or by as an unidentified mechanism, were also suggested by Fulda et al., (2004). Because of the higher pH in perioxisomes as compared to the cytoplasm, weak acids are predicted to be trapped in perioxisomes as the anions, and this may account for some of the OPDA import (Theodoulou et al., 2005).

Reduction of OPDA by OPR in Perioxisomes

Inside the perioxisomes OPDA reductase (OPR) catalyzes the subsequent steps in JA biosynthesis i.e. the reduction of the cyclopentenone ring of 9S, 13S-OPDA and dnOPDA to 3-oxo-2-(2' (Z)-pentenyl)-cyclopentane-1-octanoic acid (OPC-8:0) and hexanoic (OPC-6:0) acids, respectively (Figure 2). The signaling properties of jasmonates bearing a cyclopentenone ring (OPDA, dnOPDA) are clearly distinct from JA and its derivatives which are characterized by the reduced cyclopentanone ring (Blechert et al. 1999; Stintzi et al. 2001; Taki et al. 2005). OPR3, which is the only member of a small family of related enone reductases accepting the (9S, 13S)-enantiomer of OPDA as a substrate might therefore be particularly important for the regulation of the relative levels of these two classes of signaling molecules.

The JA precursors OPDA and dnOPDA (cyclopentenones) and jasmonates (cyclopentanones) differ in bioactivity (Blechert et al. 1999; Stintzi et al. 2001; Taki et al. 2005). The reduction of the cyclopentenone ring may therefore be particularly important, as it controls the relative levels of these two classes of signaling molecules. The reaction is catalyzed by OPR3, which is the only member of a small family of related enone reductases accepting the (9S, 13S)-enantiomer of OPDA as a substrate (Schaller et al., 2000; Stintzi and Browse, 2000; Strassner et al., 2002).

Even though OPR3 belongs to a small family of related flavin-dependent oxidoreductases (at least three in tomato, six genes in *Arabidopsis*, six in pea, eight in maize, ten in rice) it is functionally unique, as evident from JA-deficiency symptoms (male-sterility) of the *OPR3* loss-of-function mutant in *Arabidopsis* (Sanders et al., 2000; Stintzi and Browse, 2000). The genetic evidence is consistent with biochemical and structural data which suggest that all OPRs catalyze the reduction of α, β-unsaturated carbonyls (conjugated enones) while only OPR3s and the rice ortholog OsOPR7 (Tani et al. 2007) are capable of reducing the 9S,13S enantiomer of OPDA (Schaller et al., 2000; Breithaupt et al., 2001, 2006; Strassner et al., 2002). Therefore, these are the only OPRs contributing to JA biosynthesis.

The shortening of the hexanoic and octanoic acid side chains in OPC-6:0 and OPC-8:0 to yield JA involves two or three rounds of β-oxidation, respectively. Prior to entry into the β-oxidation cycle, the carboxylic moiety needs to be activated as CoA ester. In Arabidopsis thaliana, there is a large family of 63 ATP-dependent acyl-activating enzymes potentially involved in this process (Shockey et al., 2003). Within this family, a subgroup of fatty acyl-CoA synthetase was shown to activate JA precursors in vitro, including OPDA, dnOPDA, OPC-8:0, and OPC-6:0 (Kienow et al., 2008; Koo et al., 2006; Schneider et al., 2005).

However, a physiological role in JA biosynthesis was confirmed for only one of them, OPCL1 (OPC-8: CoA ligase 1, locus At1g20510) (Koo et al., 2006). Loss-of-function mutants for OPCL1 hyper-accumulate OPC-8:0, OPC-6:0, and OPC-4:0 suggesting a partial metabolic block in OPC-CoA ester formation (Kienow et al., 2008). The mutants are also compromised in wound-induced JA accumulation which is consistent with a role of OPCL1 in JA biosynthesis.

The available data are consistent with the view that OPCs are activated as CoA esters within peroxisomes and subsequently channeled into b-oxidation. The formation of OPDA-CoA and dnOPDA-CoA either in the cytosol or within peroxisomes prior to the reduction by OPR3 remains an alternative possibility.

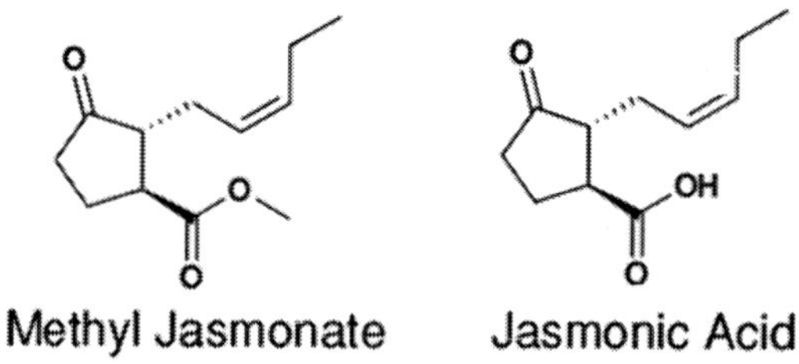

Figure 3. Structure of Jasmonic acid ($C_{12}H_{18}O_3$) and its derivative methyl jasmonate ($C_{13}H_{20}O_3$).

Beta-Oxidation

Beta-oxidation in lipid metabolism was believed to be located in the peroxisomes of all higher plants but also detected in mitochondria in a non-oilseed plant (Masterson and Wood, 2001). The terminal steps of peroxisomal JA biosynthesis are three β- oxidation reactions, which shorten the carboxyl side chain from the intermediates OPC-8:0 or OPC-6:0 produced from OPDA or dn-OPDA (Vick and Zimmernan, 1983).

Beta-oxidation itself involves three core enzymes, acyl-CoA oxidase (ACX), multifunctional protein (MFP; comprising enoyl-CoA hydratase and β-hydroxyacyl- CoA dehydrogenase activities), and 3-ketoacyl-CoA thiolase (Figure 4). ACX1A was shown to catalyze the first step in the β-oxidation of OPC-8:0-CoA, and was found to be responsible for the bulk of wound-induced JA production in tomato (Li et al., 2005).

Consistent with its essential role in JA biosynthesis, the ACX1 tomato mutant is impaired in wound-induced defense gene activation and insect resistance (Li et al., 2005). In Arabidopsis, ACX1 is responsible for about 80% of JA production after wounding (Cruz Castillo et al., 2004; Schilmiller et al., 2007), and only the ACX1/5 double mutant showed severe JA deficiency symptoms (Schilmiller et al., 2007). For MFP roles in JA biosynthesis Arabidopsis *aim1* mutant, disrupted in one of two MFP genes is similarly impaired in wound-induced accumulation of JA and the expression of JA-dependent genes (Delker et al., 2007). In tobacco a similar result was obtained, that is, a stress responsive MFP orthologous to *Arabidopsis* AIM1 are involved in β-oxidation (Ohya et al., 2008). Among five *KAT* genes in *Arabidopsis*, *KAT2* was shown to play a major role in driving wound-activated responses by participating in the biosynthesis of JA in wounded leaves (Cruz Castillo et al., 2004). In transgenic plants silenced for KAT2 expression, a reduction in wound-induced JA accumulation and JA-dependent marker gene (JR2) expression was observed (Cruz Castillo et al., 2004).

The final step of JA biosynthesis is that jasmonyl-CoA releases free acid by jasmonyl-thioesterase. There is no report of cloning of jasmonyl-thioesterase in plant so far except that in *Arabidopsis* two peroxisomal acyl-thioesterases, ACH1 and ACH2 have showed thioesterase activity of hydrolyzing both medium and long-chain fatty acyl-CoAs but not jasmonyl-CoA (Tilton et al., 2004). It becomes clear from the above discussion that of all the steps involved; only those catalyzed by AOS, AOC, and OPR3 are truly committed to the JA biosynthetic pathway.

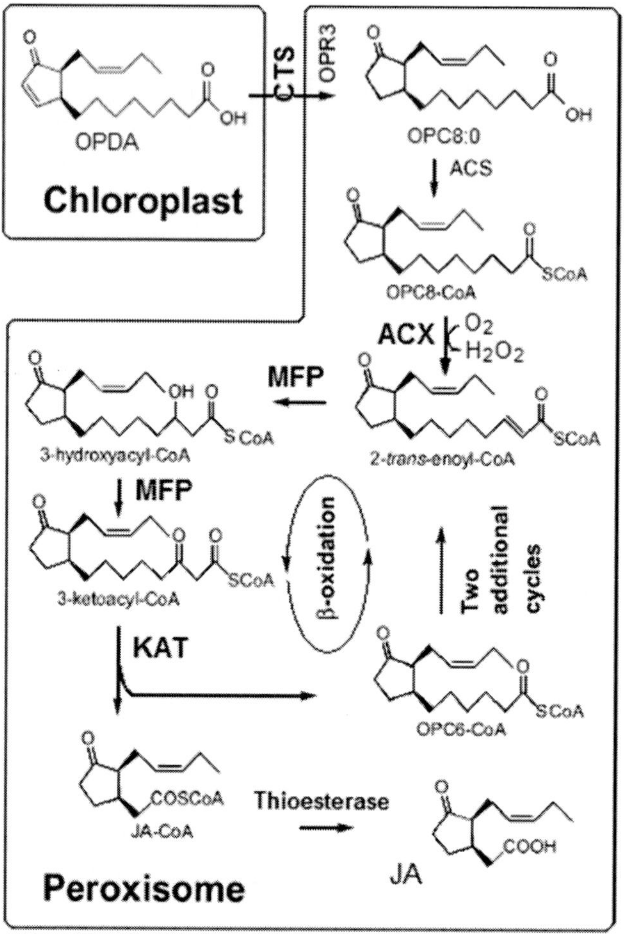

Source: Li et al., 2005.

Figure 4. β-Oxidation Scheme of JA Biosynthesis. Abbreviations: ACS (acyl-CoA synthetase), ACX (acyl-CoA oxidase), MFP (multifunctional protein), KAT (3-ketoacyl-CoA thiolase), and OPC8:0 (3-oxo-2(2' [Z]-pentenyl)-cyclopentane-1-octanoic acid).

The Enzymes for JA biosynthesis and Derivation

The enzymes of JA biosynthesis and metabolism have been extensively investigated with their reviewed structures, biochemical activities, and functional regulation (Wasternack, 2007; Schaller and Stintzi, 2009; Delker et al., 2006; Kombrink, 2012).

Phospholipase A (PLA)

Phospholipase initiates the JA biosynthesis by releasing 18:3 from chloroplast membrane galactolipids (Vick and Zimmerman, 1983). The lipase belongs to one of the following five enzyme (Wasternack, 2007; Delker et al., 2006): (1) phospholipase A1 (PLA1), which cleaves the acyl group of phospholipid and glycerolipids at the *sn*-1 position; (2) phospholipase A2 (PLA2), which cleaves the acyl group in *sn*-2 position; (3) patatin-like acyl hydrolases, which has little *sn-1/sn-2* specificity and is homologous to animal Ca2+-independent PLA2; (4) DAD-like lipase with activity of phospholipid and galactolipid acyl hydrolase that may have *sn-1* or *sn-2* specificity; (5) SAG(senescence-associated gene) 101-like acyl hyrolase.

The free hexadecatrienoic acid (16:3) liberated by a phospholipase A2 from the *sn-2* position of MGDG or DGDG can also become substrate for JA biosynthesis to form dn-OPDA (Figure 2). Alternatively, 13-LOX may oxygenate 18:3 and 16:3 esterified in galactolipids (Buseman et al., 2006; Kourtchenko et al., 2007). The resulting hydroperoxy galactolipids may then be substrates for AOS and AOC, yielding arabidopsides, which belong to galactolipid species containing esterified OPDA and dnOPDA in plastidic membranes. These lipid species may serve as storage lipids that may allow the rapid release of OPDA and dnOPDA, the JA biosynthetic intermediates. Conclusively, lipase activity is essential to supply the JA biosynthesis pathway with either free PUFA (18:3 and 16:3) or OPDA/dnOPDA as precursors.

The first reported lipase involved in JA biosynthesis was DEFECTIVE IN ANTHER DEHISCENCE1 (DAD1), a chloroplastic glycerolipid lipase. DAD1 belongs to phospholipase A1 (PLA1) family. *Arabidopsis dad1* T-DNA insertion mutants are male sterile from decreased JA accumulation required for reproduction (Ishiguro et al., 2001), but remain capable of synthesizing JA, indicating that other lipases contribute to JA production. DONGLE (DGL), a PLA1 from A. thaliana, was thought to be involved in wound-induced and basal JA biosynthesis (Yang et al., 2007; Hyun et al., 2008).

More recently, DAD1 and DGL RNAi lines were generated, and these lines were similar to the wild-type in the early wound response. The DGL protein was detected in lipid bodies but not in plastids as required for JA biosynthesis (Ellinger et al., 2010), suggesting that both enzymes are not involved in JA biosynthesis. Of an additional 16 lipase mutants screened, only PLA1γ1 (At1g06800) had a reduced level of JA in wounded leaves. However, there might still be unidentified lipases involved in wound- and pathogen-induced JA formation (Ellinger et al., 2010).

Lipoxygenase (LOX)

Lipoxygenases (LOXs) are non-heme iron-containing dioxygenases widely distributed in yeast, algae, fungus, plant, and animal species (Shin et al, 2008). Oxygenation of a-LeA is the initial step in JA biosynthesis. The oxygen has to be inserted in the C-13 position by a lipoxygenase (LOX) (Figure 1). LOX isozymes catalyze the incorporation of molecular oxygen at either position 9 or 13 of polyunsaturated fatty acids (PUFA) such as linoleic (LA,18:2) and α-linolenic (LeA,18:3) acid to produce PUFA hydroperoxides (HPOT and HPOD), which can be further converted to different oxidized fatty acids (oxylipins) through the action of enzymes participating in seven LOX-pathway branches (Figure 1). Plant LOXs are classified with respect to their positional specificity of LA dioxygenation. LOXs adding O_2 to C-9 or C-13 of the hydrocarbon backbone of LA are designated as 9-LOX or 13-LOX (Feussner and Wasternack, 2002). However, in plants some LOXs have dual positional specificity to C-9 and C-13 of LA and produce both 9- and 13-hydroperoxides of linoleic acid (Hughes et al. 2001, Kim et al. 2002, Garbe et al. 2006).

Among the six LOXs of Arabidopsis, four of them are 13-LOXs (LOX2, LOX3, LOX4, LOX6) (Bannenberg et al., 2009), although their functions are only partly understood. LOX2 was thought to be involved in the wound response for a long time (Bell et al., 1995) and subsequent studies revealed that LOX2 was responsible for the bulk of JA formation in the first h upon wounding (Glauser et al., 2009; Schommer et al., 2008)

Plants LOXs were found in several subcellular compartments including chloroplast, vacuole and cytosol. According to their localization and sequence similarity, plant LOXs can be classified into type 1- and type 2-LOXs. Type 1-LOXs harbor no chloroplast-transit peptide but the members of this group share a high similarity (>75%) of amino acid sequence to one another. Type 2-LOXs carry a putative chloroplast transit peptide and show only a moderate overall similarity (35%) of amino acid sequence to one another. To date, these LOX forms all belong to the subfamily of 13-LOXs (Feussner and Wasternack, 2002). After LOX activity and fluxing through the lipoxygenase pathway branches, PUFA (mainly LA and LeA) can be converted to hundreds of oxylipin species which physiological roles are largely unclear in plants.

Every plant species harbors several LOX isozymes, encoded by a LOX gene family. For example, the *Arabidopsis* genome contains six LOX genes (Bannenberg et al., 2009), while rice and maize have 14 (Umate, 2011) and 13 LOX genes (Nemchenko et al., 2006), respectively. All LOX isoforms may contribute to oxylipin production, but only 13-LOXs with chloroplast-transit peptide participate in Vick and Zimmerman pathway for JA production. Additionally, several LOX genes may function for JA biosynthesis, e.g., LOX2, LOX3, LOX4 and LOX6 in *Arabidopsis* contain chloroplast signaling peptides and show 13S-lipoxygenase activity, both required for JA biosynthesis (Bannenberg et al., 2009). Plant LOXs have been also correlated with seed germination, vegetative and reproductive growth, fruit maturation, plant senescence, and responses to pathogen attacks and insect wounding (Porta and Rocha-Sosa, 2002).

Allene Oxide Synthase (AOS)

Allene oxide synthase (AOS) catalyzes 9-/13-HPOT to the unstable epoxide, either 9,10- EOT (9,10-Epoxyoctadecatrienoic acid) or 12,13-EOT (12,13-Epoxyoctadecatrienoic acid) (Figure 1). The unstable epoxide can be either hydrolysed non-enzymatically to α- and γ -ketols or cyclized to 12-oxo-phytodienoic acid (OPDA). Only 12, 13-EOT provides substrate for the following enzyme in JA biosynthesis, allene oxide cyclase (AOC) (Figure 1).

AOSs belong to an atypical cytochrome P450 subfamily, the CYP74 enzymes, which are independent from molecular oxygen and NADPH, exhibit low affinity to CO, and use an acyl hydroperoxide as the substrate and the oxygen donor (Stumpe and Feussner, 2006). According to the specificity of AOS to the substrates, 9-/13- hydroperoxides, AOS enzymes specialize into 9- or 13-AOS, which use either 9- or 13- hydroperoxide, respectively, as substrate. AOS enzymes from barley and rice show no substrate specificity for either (9*S*)-hydroperoxides or (13*S*)-hydroperoxides, and designated 9/13-AOS (Stumpe and Feussner, 2006). CYP74 enzymes rearrange their fatty acid hydroperoxide substrates into various structurally different products:

13-AOSs catalyze the formation of 12,13-(S)-epoxy-octadecanoic acid and 10,11-(S)-epoxy hexadecatrienoic acid (10,11-EHT) as the first committed step of JA biosynthesis, whereas divinyl ether synthases (DESs) convert hydroperoxides to divinyl ethers, and hydroperoxide lyases (HPLs) produce short-lived hemiacetals that decompose to aldehydes and x-fatty acids (Howe and Schilmiller, 2002; Stumpe and Feussner, 2006)

Like LOXs, only 13-AOS functions in JA biosynthesis. All 13-AOS carry a plastid-transit peptide except AOS from guayule (Pan et al., 1995) and barley (Maucher at al., 2000), indicating that during JA biosynthesis, AOS localizes to chloroplast. Plant species may contain one or multiple AOS genes. For example, *Arabidopsis* jus has a single copy of AOS gene while rice may have four AOS genes (Agrawal et al., 2004). AOS genes from plant species such as flax (Song et al., 1993), guayule (Pan et al., 1995), *Arabidopsis* (Laudert et al., 1996), tomato (Howe et al., 2000), barley (Maucher et al., 2000), rice (Ha et al., 2002; Agrawal et al., 2004) and corn (Utsunomiya et al., 2000) have been cloned or purified so far.

Overexpression of flax AOS in transgenic potato plants led to 6-12 folds increased of basal JA level (Harms et al., 1995). However, overexpression of *Arabidopsis* AOS in either *Arabidopsis* or tobacco did not alter the basal level of JA (Laudert et al., 2000), indicating that the basal expression level of AOS varied in plant species which may be the bottleneck or not for JA production in rest plants. One important property of *AOS* genes in plants is that they are strongly induced by wounding and JA- / MeJA-, and OPDA- treatment in many plant species (Harms et al., 1995; Laudert and weiler, 1998).

Allene Oxide Cyclase (AOC)

Allene oxide cyclase (AOC) catalyzes the stereospecific cyclization of the unstable allene oxide, the product of AOS into the cis-(+) enantiomer OPDA, the precursor of JA (Figure 2). The unstable allene oxide is either 9,10-EOT (9,10-Epoxyoctadecatrienoic acid) or 12,13-EOT (12,13-Epoxyoctadecatrienoic acid), corresponding to 9-/13-HPOT, the substrates of AOS. These unstable substrates of AOC, 9,10-EOT and 12,13-EOT can spontaneously and rapidly hydrolyze to a mixture of α- and γ- ketols ($t1/2 < 30$ minutes in water) (Schaller et al., 2004). In presence of AOC, the cis (+)-isomers of OPDA and dnOPDA are formed exclusively as precursors of bioactive JAs. AOC is thus crucially important to establish the enantiomeric structure of the cyclopentenone ring. However, *in vivo* α- and γ- ketols are not detectable (Schaller et al., 2004), suggesting tight coupling of AOS and AOC reactions, which effectively convert HPOT into OPDA.

The enzyme was first isolated from corn seeds (Ziegler et al., 1997) and a few years later cloned from tomato. AOC is encoded by a single gene in tomato and a small gene family in Arabidopsis (Stenzel et al., 2003b). Despite comprehensive analyses of substrate specificity (Ziegler et al., 1999), the mechanism by which stereochemical control is exerted on the cyclization reaction remained enigmatic until the recent elucidation of several crystal structures for Arabidopsis AOCs (Hofmann and Pollmann, 2008; Hofmann et al., 2006; Levin et al., 2007).

Recently, differential expressions of the four AOCs of A. thaliana were demonstrated by corresponding promoter: GUS lines (Stenzel et al., 2012). Recently, the crystal structure of AOC1 and AOC2 of Physcomitrella patens revealed new mechanistic insights into AOC catalysis, including tight binding of the substrate, accompanied by conformational changes within the binding pocket (Neumann et al., 2012). AOC and other enzymes in JA biosynthesis such as LOX and ALLENE OXIDE SYNHASE (AOS) are partially associated with chloroplast membranes (Farmaki et al., 2007). The association of LOX, AOS and AOC with chloroplast membranes implies that the esterified OPDAs, called arabidopsides, may be formed from fatty acids esterified in galactolipids by membrane-bound enzymes, as indicated by recent labeling experiments (Nilsson et al., 2012).

Oxo-Phytodienoic Acid Reductase (OPR)

The first step of peroxisomal JA biosynthesis is the conversion of OPDA, a cyclopentenone to cyclopentanone (3-oxo-2-(2' (*Z*)-pentenyl)-cyclopentane-1-octanoic acid, OPC-8:0) catalyzed by OPDA reductase (OPR). OPR enzymes belong to small family of related flavin-dependent oxidoreductases (Old Yellow Enzyme-OYE) (EC 1.6.99.1), initially isolated from brewer's bottom yeast and shown to possess a flavin Cofactor. The first identified OYE in higher plant is OPR1 of *Arabidopsis* (Schaller and Weiler, 1997). Plant OPR isomers are encoded by small gene families identified in across broad plant genre, (Schaller et al., 2004). Better-studied OPR families include five *OPRs* (three of which are characterized) in *Arabidopsis* (Sanders et al., 2000), six in pea (Matsui et al., 2004), three in tomato (Strassner et al., 2002), 13 in rice (Agrawal et al., 2004), and eight in maize (Zhang et al., 2005). All theses OPRs can catalyze the reduction of α, β-unsaturated carbonyls (conjugated enones) in a wide spectrum of substrates in including four stereoisomers of OPDA (Sanders et al., 2000).

Earlier studies on the enzymatic activity of OPRs in *Arabidopsis* and tomato revealed that different OPR isomers have distinct substrate preferences to warrant classification into separate groups, group I and II, depending on their substrate specificity to OPDA stereoisomers (Schaller et al., 1998). OPR group I enzymes preferentially catalyze the reduction of (9*R*,13*R*)-12-oxo 10,15(*Z*)-octadecatrienoic acid (9*R*,13*R*-OPDA), while OPR group II enzymes preferentially catalyze (9*S*,13*S*)-12-oxo-10,15(*Z*)-octadecatrienoic acid (9*S*,13*S*-OPDA), a intermediate biosynthetic precursor in JA biosynthesis

(Schaller et al., 1998). OPR3 of *Arabidopsis* and tomato, belonging to group II, have been shown to efficiently reduce the natural isomer 9*S*, 13*S*-OPDA to OPC 8:0, the precursor of JA (Schaller et al., 1998).

In addition, to uncover the molecular determinants of substrate specificity between OPR group I and II, crystal structural comparison and mutational analysis of tomato OPR1/3 in complex with OPDA enantiomers revealed that two active-site residues, i.e., Tyr78 and Tyr246 in OPR1 and Phe74 and His244 in OPR3 of tomato, are critical for substrate specificity (Breithaupt et al. 2009). Among the six OPRs of A. thaliana only OPR3 is involved in JA biosynthesis (Wasternack and Kombrink, 2010). The unique function of OPR3 in jasmonate biosynthesis is a result of its relaxed stereoselectivity with respect to OPDA stereoisomers. The physiologically relevant (9S,13S)-enantiomer of OPDA is not a substrate for the majority of OPRs, but is reduced exclusively by OPR3 from tomato and Arabidopsis, and orthologs in other species (OsOPR7) (Breithaupt et al., 2001, 2006; Jung et al., 2007). Thus, the biochemical studies conclude OPR3 rather than OPR1 and OPR2, is responsible for JA production in *Arabidopsis* and tomato.

Genetic studies on a knockout *Arabidopsis* mutant, *dde1/opr3,* displayed male sterilility and compromised defense responses resulting from JA deficiency, indicating OPR3 is essential for the JA biosynthesis pathway and other OPR isomers such as OPR1/2 cannot substitute for OPR3 in JA production (Sanders et al., 2000; Stintzi and Browse, 2000). *opr7 opr8* double mutant showed a number of genetic phenotypes such as tassel seed and susceptibility to insect and pathogen, reflecting JA essential functions in monocotyledonous plants (Yan et al., 2012). Thus in plants, the physiological role of OPR group II enzymes in plants is primarily for production of the jasmonates, which mediate many development and defense-related processes (Yan et al., 2012).

Carboxyl Methyltransferase (JMT)

The floral scent methyl jasmonate (MeJA) has been identified as a vital cellular regulator that mediates diverse developmental processes and defense responses against biotic and abiotic stresses (Cheong and Choi, 2003). Apart from its putative roles in defense responses, MSA, with its characteristic wintergreen scent, is an important chemoattractant for moth-pollinated flowering plants, including Clarkia (Raguso and Pichersky, 1995; Raguso et al., 1996; Dudareva et al., 1998). The enzyme converting JA to methyl jasmonates (MeJA) is JA carboxyl methyltransferase (JMT), which was first cloned and characterized from *Arabidopsis* (Seo et al., 2001). As JMT does not carry any transit signal peptides, it is presumably a cytoplasmic enzyme. JMT is constitutively expressed in almost all the organs of mature plants, but not in young seedling (Seo et al., 2001), indicating young plant avoids to produce MeJA. However, JMT can be induced by both wounding and MeJA treatment (Seo et al., 2001). Transgenic *Arabidopsis* over expressing *JMT* exhibited constitutive expression of jasmonate-responsive genes, including *VSP* and *PDF1.2*, and enhanced level of resistance against the virulent fungus *Botrytis cinerea*, indicating JMT is a key enzyme for airborne-jasmonate-regulated plant responses.

JA-Amino Acid Synthetase (JAR1)

JAR1 (JASMONATE RESISTANT 1) is a jasmonate amino acid transferase that plays an important role in jasmonic acid metabolism and jasmonic acid signaling (Guranowski et al., 2007, Suza and Staswick, 2008). JA amino acid synthetase conjugates isoleucine to JA (Staswick and Tiryaki, 2004) to form jasmonoyl-L-isoleucine (JA-Ile), the bioactive ligand involved in JA signaling.

JAR1 is one of 19 closely related *Arabidopsis* genes that are similar to the auxin-induced soybean *GH3* gene family (Staswick et al., 2002). Analysis of fold-predictions for this protein family suggested that JAR1 might belong to the acyl adenylate-forming firefly luciferase superfamily. These enzymes activate the carboxyl groups of a variety of substrates including JA, indole-3-acetic acid (IAA) and salicylic acid (SA) for their subsequent biochemical modification (Staswick et al., 2002), thereby regulating hormone activity. The first *jar1* mutant was identified that affected signaling in the jasmonate pathway. *Jar1* plants have reduced sensitivity to root growth inhibition in the presence of exogenous JA (Staswick et al., 2002). In wounding *JAR1* transcript was found increased dramatically in wounded tissue and JA–Ile accumulated mostly near the wound site with a minor increase in unwounded tissue (Suza and Staswick, 2008). However, the reduced accumulation of JA–Ile had little or no effect on several jasmonate-dependent wound-induced genes such as *VSP2,* for *LOX2*, *PDF1.2*, *WRKY33*, *TAT3* and *CORI3*. Morphologically, *jar1* mutation is male fertile while JA biosynthesis and signaling mutants are male sterile (Suza and Staswick, 2008).

In vivo, JAR1 appears to be primarily involved in the production of jasmonoyl-isoleucine (JA-Ile). This compound has been shown to be an active form of jasmonic acid that promotes the interaction between COI1 (an F-box) protein and the JAZ1 repressor protein in *Arabidopsis thaliana*. In this manner, JA-Ile is believed to help destabilize the JAZ1 repressor and thereby promote JA-responsive transcription. *In vitro*, JAR1 can catalyze the addition of many different amino acids to jasmonate (Staswick and Tiryaki, 2004; Guranowski et al., 2007; Suza and Staswick, 2008). A few of these JA-amino acid conjugates have been identified in Arabidopsis thaliana, including JA-valine, JA-leucine, JA-phenylalanine, and JA-ACC, but their levels do not change in a jar1 mutant suggesting that other enzymes must catalyze the formation of these products *in vivo* (Staswick and Tiryaki, 2004). *In vitro* analyses of JAR1 show that its Km for isoleucine (0.03 mM) is more than 60-fold lower than its Km for leucine (1.93 mM), phenylalanine (1.84 mM), and valine (2.49 mM) and provide an explanation for the *in vivo* production of JA-Ile by this enzyme.

In addition, although JAR1 can act on (+)-jasmonate and OPC-4 (3-oxo-2-(2Z-pentenyl)-cyclopentane-1-butyric acid) *in vitro*, it shows the highest rate of activity with (-)-jasmonate (Staswick and Tiryaki. 2004). Several *jar1* mutants have been identified and studied (Suza ans Staswick, 2008). These mutants demonstrate that JAR1 activity is required for wounding responses, protection against ozone damage, resistance to fungi and pathogens, and other jasmonate-related responses.

Derivatives and Metabolites of JA

Traditionally, JA has been viewed as the end product of the pathway and the bioactive hormone. The many different metabolites of JA and the enzymatic steps in their interconversion were thought to contribute to hormone homeostasis to sustain and control the levels of active JA. This view is changing, however, and E.E. Farmer takes the extreme position to consider JA as a mere precursor of bioactive hormones, i.e. JA conjugates (Farmer, 2007).

The JA biosynthetic pathway from linolenic acid yields (+) -7-iso-JA (3R, 7S-JA) as the final product (Sembdner and Parthier, 1993). However, this molecule readily isomerizes to the thermodynamically favoured stereoisomer (-)-JA (3R, 7R-JA) resulting in a molar equilibrium of about 9: 1 ((-)-JA: (+)-7-iso-JA) under normal conditions (Sembdner and Parthier, 1993).

In addition to isomerization, JA undergoes a series of molecular modifications to form a variety of metabolites in plants (Figure 5).

1. C1 carboxyl group can be methyl-esterified or conjugated with amino acids or with 1-Aminocyclopropane-1-carboxylic acid (ACC).
2. C1 carboxyl group can be decarboxylated, 3) Glycosylation of C1 carboxyl group.
3. Reduction of C6 carbonyl group, 5) Reduction of C9, 10 double bond.
4. Hydroxylation of carbon at C11 or C12.

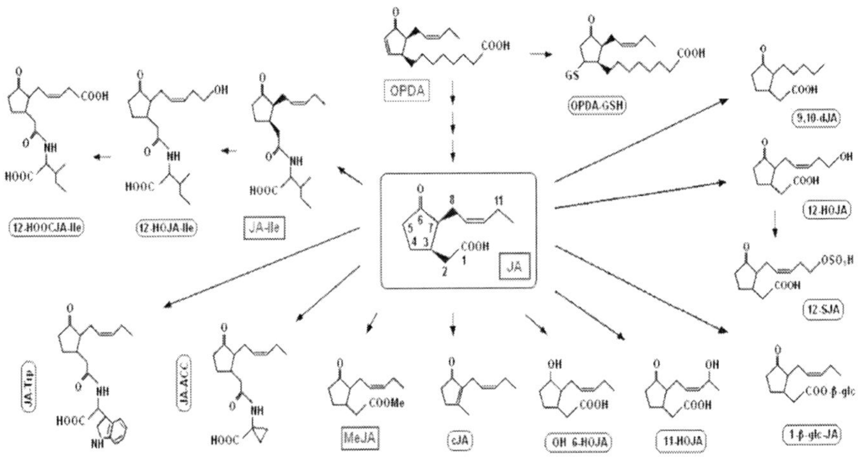

Source: Gfeller et al., 2010.

Figure 5. The metabolites produced from JA in plants (Yuanxin Yan et al., 2013).

With the above reactions, (+) -7-iso-JA can be converted into more than 30 distinct jasmonates which were found to be widespread in Angiospermae, Gymnospermae, Pteridophyta, Algae such as Euglena, Spirulina, and Chiarella, and the red alga Gelidium (Sembdner and Parthier, 1993). However, just several jasmonates e.g., free JA, cis-jasmone, MeJA and JA-Ile, are considered to be the major bioactive JA forms in plants (Fonseca et al., 2009). Other jasmonate derivatives or conjugates have been viewed as clearance metabolites playing important roles in hormone homeostasis (Sembdner and Parthier, 1993), in which JA biosynthesis and JA degradation is balanced to control the actual active JA level for fine-tuning developmental and defensive events.

Regulation of Jasmonic Acid biosynthesis

The levels of jasmonic acid in plants vary with developmental stage, organs, and are variable in response to different environmental stimuli (Creelman and Mullet, 1995). High levels of jasmonates are found in flowers, pericarp tissues of developing fruit, and in the chloroplasts of illuminated plants; Jasmonate levels increase rapidly in response to mechanical perturbations such as tendril coiling and when plants suffer wounding (Creelman and Mullet, 1995). There are several strategies applicable for plants to regulate generation or activities of jasmontes. The first strategy is to regulate the expression of the enzymes in JA biosynthesis pathway. Many studies show that most enzyme genes for JA biosynthesis such as *LOX*, *AOS*, *AOC*, *OPR3*, *JMT* and *JRA1* are induced by JA treatment (Wasternack, 2007) or wounding (Schaller, 2001), implying that JA biosynthesis is a JA dependent process in plants. Monitoring of MeJA-responsive genes in *Arabidopsis* by cDNA microarrays also concluded that JA biosynthesis is regulated by a positive feedback loop (Sasaki et al., 2001). Further evidence for this conclusion has come from mutants with constitutively up-regulated JA levels such as *cev1* and *fou2*. These mutants showed typical phenotypes associated with exogenous JA treatment such as roots/shoots growth inhibition, anthocyanin accumulation in the leaves, and over-expression of JA-dependent genes (Ellis and Turner, 2001; Bonaventure et al., 2007). However, the expression levels of JA biosynthetic enzymes do not determine the actual output of JA biosynthesis in limited substrate conditions.

The second approach for JA biosynthesis regulation is controlling the substrate availability. For example, the fully expanded leaves of *Arabidopsis* carry LOX, AOS, and AOC proteins abundantly; however, JA formation is at a substantially low level. JA production rapidly occurs only upon strong external stimuli such as wounding, which largely induces JA biosynthesis enzymes and releases the substrate LA from the membranes (Stenzel et al., 2003). Furthermore, wound induction of JA is transient and appears before the expression induction of *LOX*, *AOS*, and *AOC* genes (Howe et al., 2000). These data clearly show that JA biosynthesis is regulated by enzyme activities and substrate availability.

The third strategy of JA biosynthesis regulation is to store or reuse intermediates or conjugates of jasmonate in case of over-produced or quick release is required for healing in situation of insect or pathogen attacks. Recently, esterified OPDA was found in galactolipids (monogalactosyldiacylglycerol, MGDG and digalactosyldiacylglycerol,

DGDG) (Stelmach et al., 2001). A novel oxylipin category, so called arabidopsides A, B, C, D, E, F, G and F (Figure 5) were found containing OPDA and/or dino-OPDA (Hisamatsu et al., 2003; Hisamatsu, 2005). These compounds could accumulate to 7-8% of total lipids in plants if challenged by pathogens (Anderson et al., 2006). In addition, JA derivatives including jasmonyl-amino acids such as JA-Ile (jasmonoylisoleucine), JA-Leu (jasmonoyl-leucine), and JA-Val (jasmonoyl-valine) and jasmonyl-ACC (1-amino- cyclopropane-1-carboxylic acid) conjugates are present and all, except JA-Ile, are considered storage forms of jasmonates in plants (Staswick et al., 2002).

The biosynthesis of JA-Ile is tightly controlled by developmental and environmental cues (Creelman and Mullet, 1997; Wasternack, 2007; Koo and Howe, 2009). Consistent with their role in regulating induced defenses in vegetative tissues, JA and JA-Ile typically accumulate in response to various biotic and abiotic stresses. Mechanical wounding of *Arabidopsis* leaves, for example, effectively triggers *de novo* accumulation of JA/JA-Ile within minutes of tissue damage (Chung et al., 2008; Glauser et al., 2008; Suza and Staswick, 2008; Koo and Howe, 2009; Koo et al., 2009; Suza et al., 2010). The rapidity of this response indicates that all JA-Ile biosynthetic enzymes, including JAR1, are present in unstressed cells prior to stimulation.

Genes encoding many of the enzymes involved in JA-Ile biosynthesis are coordinately up-regulated in response to environmental signals that activate production of the hormone (Reymond et al., 2000; Sasaki et al., 2001; Sasaki-Sekimoto et al., 2005; Koo et al., 2006; Ralph et al., 2006; Pauwels et al., 2008). This transcriptional response presumably constitutes a positive feedback mechanism to amplify the cell's capacity for JA metabolism. At a practical level, this co-expression phenomenon has proven useful for the identification of novel components in the JA metabolic and signaling pathways (Koo et al., 2006, 2011; Thines et al., 2007; Heitz et al., 2012). The ability of exogenous JA and MeJA to potently activate hormonal responses can be attributed to their *in vivo* conversion to JA-Ile and subsequent action through the COI1–JAZ receptor system (Tamogami et al., 2008; Wu et al., 2008).

Ca2+ and MAPK cascades are also involved in the regulation of JA biosynthesis (Wasternack and Hause, 2013). In *A. thaliana*, MKK3 and MPK6 are activated by JA leading to negative regulation of MYC2 expression and repression of JA biosynthesis genes (Takahashi et al., 2007). In a parallel pathway, however, there is an MKK3/MPK6-independent activation of MYC2 by JA, and the MKK3/MPK6 cascade is epistatic to MYC2 (Takahashi et al., 2007). There exists a link between JA biosynthesis and MAPK pathways, as revealed by co-expression analyses of microarray datasets in *A. thaliana* (Van Verk et al., 2011). Here, it became obvious that OPR3 and to a minor extent AOS are co-expressed with MYC2, MEK1, MEKK1, MKK4 and MPK3.

Ca^{2+} is a universal second messenger, which is activated very early in signaling cascades and holds an important place in plant signaling as a mediator for response against a wide array of biotic and abiotic stimuli (Kudla et al., 2010). Ca^{2+} is a toxic cellular compound at high concentrations because of its ability to form insoluble complexes with proteins, membranes, and organic acids. So a tight spatial and temporal control over Ca^{2+} concentration is maintained in a cell, with the cytosol having a lower concentration than the external medium and subcellular compartments, resulting in a steep gradient. Upon perception of a signal, Ca^{2+} elevations arise in the cytosol due to an influx of Ca2+ from different stores, and this subsequently activates signaling (Dodd et al., 2010). Ca^{2+} is an early acting second messenger in response to many biotic and abiotic stimuli (Kudla et al., 2010).

Arabidopsis has at least 250 proteins with predicted Ca2+ binding EF hand domains. Such a large number of genes involved in a single pathway point to the importance of Ca2+ signaling in biological responses. The importance of Ca2+ has also been demonstrated in experiments with *P. lunatus*, which in response to attack by the spider mite *Tetranychus urticae* shows activated defense gene expression requiring Ca2+ influx and protein phosphorylation (Arimura et al., 2000).

In Arabidopsis, the calmodulinbinding protein IQD1 was shown to be involved in plant defense against feeding by the aphid *Myzus persicae* and the *lepidopteran Trichoplusia* by affecting glucosinolate (GS) biosynthesis (Levy et al., 2005) Ca^{2+} has also been implicated in downstream parts of herbivore triggered signaling cascades, because jasmonates and synthetic analogs induce Ca2+ elevations in tobacco (*Nicotiana benthamiana*) BY-2 cell culture (Walter et al., 2007).

For wound and herbivore-induced JA accumulation in *Nicotiana attenuata*, the Ca^{2+}-dependent protein kinases CDPK4 and CDPK5 are negative regulators (Yang et al., 2012), whereas a wound-induced protein kinase (WIPK) is rapidly activated near the wound region thereby activating JA biosynthesis (Wu et al., 2007).

In tomato, a MPK1, MPK2 and MPK3 are involved in expression of JA biosynthesis genes (Kandoth et al., 2007). Here, the activation of MPKs is systemin-dependent. Further control of JA biosynthesis is mediated by the COP9 (CONSTITUTIVE PHOTOMORPHOGENESIS 9) signalosome (CSN), a multiprotein complex involved in the regulation of CULLIN-RING E3 ubiquitin ligases.

CSN not only is required for optimum plant development, but is also involved in plant defence against herbivores and pathogens by its modulation of JA levels (Hind et al., 2011).

In *A. thaliana*, the overexpression of a plasma membrane located glutamate receptor results in increased glutamate mediated Ca2+ influx and resistance to necrotrophic pathogens (Kang et al., 2006). A putative link to JA biosynthesis is lacking, but is suggested by the up-regulation of VSP1, LOX2 and other JA-responsive genes.

JA Perception and Signaling

JA signaling systems modulate plant immune responses and confer resistance against different pathogens. Activation of JA signaling system induces resistance against *Botrytis cinerea* (Moffat et al., 2012) and *Alternaria brassicicola* (Kidd et al., 2009). Supression of JA signaling system results in the suppression of root-knot nematode infection in rice (Nahar et al., 2011). Mutants of the impaired JA-perception gene coi1, but not JA biosynthesis mutants, wee highly resistant to *F.oxysporum* in *Arabidopsis thaliana* (Thatcher et al., 2009). The recent studies suggest JA signaling system is an important component in the plant immune signaling system.

The JA signaling pathway involves several Signal transduction events: the perception of the primary wound or stress stimulus and transduction of the signal locally and systemically; the perception of this signal and induction of JA biosynthesis; the perception of JA and induction of responses; and finally, integration of JA signaling with outputs from the SA, Ethylene, and other signaling pathways (Figure 6). Signaling in the Jasmonate pathway depends on at least two massive signaling machines that interact to form a complex of 0.7 mDa. The first of these Complexes is the SCF-COI1 complex, which is an E3-ubiquitin ligase. The second multi-molecular complex involved in JA signaling is the CSN (COP9 Signalosome Complex).

SCF Complex

Plants utilize the ubiquitin-proteasome system (UPS) to modulate nearly every aspect of growth and development. The ubiquitin-proteasome system is the central regulator in plant hormone sensing and signaling. Ubiquitin is covalently attached to target proteins through the action of three enzymes known as E1, E2, and E3. The SCF-complex is composed of four subunits: Cul1 (Cullin-1), Skp1, RBX1 (RING-Box Protein-1) and an F-box protein, which recognizes a target protein which is ubiquitinated and subsequently subjected to proteasomal degradation. The Cul1 requires Rub (Ubiquitin-Related Protein) modification mediated by AxR1 (Auxin Resistant-1)-Ecr1 (RUb-activating enzyme-1) for normal activity of the complex. AxR1 links the Auxin and JA signaling pathways. The SCF -complex physically associates with a protein COI1 (Coronatine-Insensitive Protein-1) to form active SCF-COI1 complex. *COI1* locus encodes an F-box protein which is known to associate with SKP1, Cullin, and Rbx proteins to form an E3 ubiquitin ligase, also known as the SCF complex ((Wasternack and Hause, 2013; Kazan and Manners, 2008).

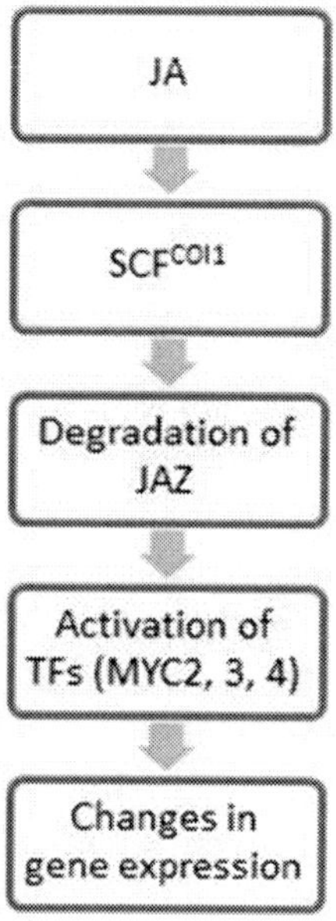

Figure 6. Major components of the jasmonate signaling pathway.

The investigators suggested that COI1, rather than JAR1 (*Jasmonates resistant 1)*, is a potential JA-receptor or a component of a receptor complex based on the study in *coi1* mutant, which displays severe JA signal-phenotypes such as male sterility, defective responses to JA-treatment and wounding, and high susceptibility to insect and necrotrophic pathogens whereas *jar1* is fertile and only partially defective to JA-treatment and wounding. The components of SCF^{COI1} complex were demonstrated to exist in Arabidopsis and mutations in the components of SCF resulted in reduced JA-dependent responses (Xu et al., 2002). The substrates of SCFCOI1 complex, is called JAZ proteins (Yan et al., 2007).

JAZ Proteins (JAZMONATE ZIM DOMAIN Proteins)

Members of the family of JAZ proteins are key regulators of the jasmonate (JA) hormonal response. JAZ proteins regulate JA-responsive gene transcription by inhibiting DNA-binding transcription factors in the absence of JA. JAZ proteins interact in a hormone-dependent manner with CORONATINE INSENSITIVE 1 (COI1), the recognition component of the E3 ubiquitin ligase, SCF^{COI1}, resulting in the ubiquitination and subsequent degradation of JAZs via the 26S proteasome pathway. Since their discovery in 2007, JAZ proteins have been implicated in protein–protein interactions with multiple transcription factors. These studies have shed light on the mechanism by which JAZs repress transcription, are targeted for degradation, modulate the JA signaling response, and participate in crosstalk with other hormone signaling pathways. In this review, we will take a close look at the recent discoveries made possible by the characterization JAZ protein–protein interactions.

There are 12 JAZ proteins in A. thaliana (Pauwels and Goossens, 2011; Wager and Browse, 2012). They contain a weakly conserved N-terminal domain, a highly conserved C-terminal Jas domain that mediates the interaction with the transcription factors, and the conserved ZIM (TIFY) domain responsible for JAZ dimerization and interaction with NINJA (Vanholme et al., 2007; Chung et al., 2009; Pauwels and Goossens, 2011). The Jas domain is exclusively required for the repressive activity of JAZ proteins (Chini et al., 2007; Thines et al., 2007; Yan et al., 2007). The expression of truncated JAZs lacking the Jas domain was associated with dominant insensitivity to exogenous JA. The initial assumption that individual JAZ proteins act specifically with different targets was subsequently revised by numerous interaction studies and the fact that there is a common occurrence of the ZIM and the Jas domain; for example, all 12 JAZs interact with MYC2, and JAZ1 interacts with nearly all target proteins mentioned above. The JA signaling is, however, mediated by a JAZ-regulatory network that entails interaction with multiple transcription factors, formation of homo- and heterodimers, alternative splicing of JAZ-encoding genes and differential stability of JAZs (Pauwels and Goossens, 2011; Kazan and Manners, 2012; Shyu et al., 2012).

JA Signaling Model: SCF^{COI1}/JAZ Proteins

Under normal growth conditions JA hormone levels are low and JA-mediated responses are kept in a repressed state by JAZ proteins. JAZs function as inhibitors of transcription factors that, when free of JAZ inhibition, act to activate transcription from JA-response genes.

In response to stresses, such as those that result from insect feeding or necrotrophic pathogen infection, an increase in levels of JA–Ile allows the hormone to facilitate interaction between the JAZ repressors and the F-box protein CORONATINE INSENSITIVE 1 (COI1), the recognition component of the E3 ubiquitin ligase SCF^{COI1}. This interaction targets JAZs for ubiquitination and degradation via the 26S proteasome pathway, releasing DNA-binding transcription factors and allowing for the induction of JA responsive genes (Figure 1) Chini et al., 2007; Thines et al., 2007). Though protein–protein interactions have been required to elucidate the JA sensing mechanism, the COI1 protein was identified through the study of mutants. One of the first JA signaling mutants, *coi1-1* (Feys et al., 1994), was identified in a screen for *Arabidopsis* seedlings resistant to root growth inhibition by coronatine, a bacterial toxin bearing structural similarity to JA–Ile, the isoleucine-conjugated derivative of JA later identified as the active form of the hormone. *coi1-1* mutants were male sterile, resistant to root growth inhibition by JA, and deficient in all JA-related responses (Feys et al., 1994; Xie et al., 1998). According to the model, in response to JA these repressor proteins would be targeted for degradation through the SCF^{COI1}/26S proteasome pathway, liberating DNA-binding transcription factors and allowing for transcription of JA-responsive genes (Figure 7) (Browse, 2009a).

Transcription factors dependent on JA signal are supposed to be the key components for JA signaling pathway. Up to date, MYC2 is the only transcription factor known to interact directly with JAZ proteins. According to the current model of JA signaling (Figure 6), MYC2 is the most important transcription factor to activate transcription of the early JA responsive genes including downstream transcription factors (such as WRKYs, MYBs, and AP2/ERFs), JA biosynthesis genes, and JAZ proteins (Lorenzo et al., 2004;

Chini et al., 2007; Chung and Howe, 2009). MYC2 was identified as a key regulator of JA signaling, acting as a basic helix-loop-helix (bHLH, also called MYC) transcription factor for JA signal transduction. MYC2 differentially regulates two branches of JA-mediated responses. That is, it positively regulates a wound-responsive gene set, including *VSP2*, *LOX3*, and *TAT*, but represses the expression of a pathogen responsive gene set such as *PR4*, *PR1*, and *PDF1.2* (Lorenzo et al., 2004).

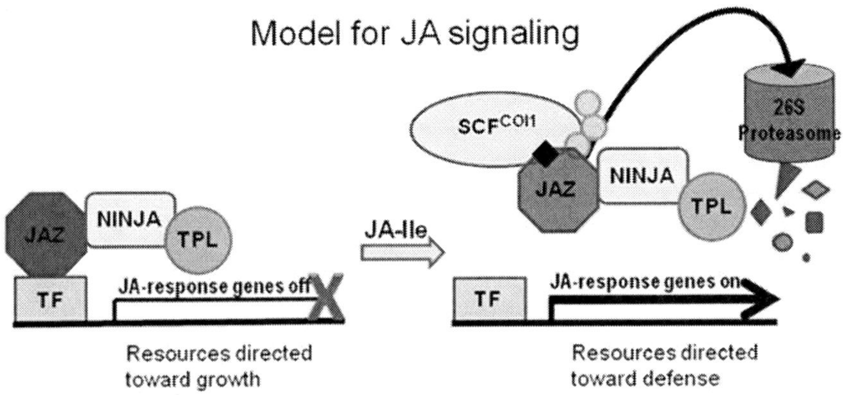

Source: Wager and Browse, 2012.

Figure 7. JAZ proteins bind to transcription factors (TFs) and recruit co-repressors, including NINJA and TOPLESS (TPL), to repress gene transcription in the absence of the active form of jasmonate (JA), JA–Ile (black diamond). In the presence of JA–Ile, the hormone facilitates interaction between JAZ proteins and the F-box protein COI1, the recognition component of the E3 ubiquitin ligase SCFCOI1. As a result, JAZ proteins are ubiquitinated (orange circles) and subsequently degraded in the 26S proteasome releasing TFs from inhibition and activating JA-responsive gene transcription.

AP2/ERF transcription factors are considered the second important group of transcription factors that belong to a large plant-specific APETALA2/ETHYLENE RESPONSE FACTOR (AP2/ERF) superfamily, containing at least 122 members in *Arabidopsis* (Nakano et al., 2006). Many ERF genes have been shown to be regulated by a variety of stress related stimuli, such as wounding, JA, ethylene, salicylic acid, or infection by different types of pathogens (Pré et al., 2008). Four transcription factors, ERF1, AtERF2, AtERF14, and ORA59, were suggested to function as positive regulators involving in JA signing pathway while AtERF4 was identified as a negative regulator for this process (Lorenzo et al., 2003; Pré et al., 2008; McGrath et al., 2005).

COI1–JAZ Co-Receptor Complex

In this complex, the Jas domain of JAZ proteins interacts with COI1, if the ligand JA-Ile is present. This interaction takes place via the N-terminal 20 amino acid residues of the Jas degron, and is strongly increased by IP5 (Inositol pentakisphosphate) (Sheard et al., 2010; Mosblech et al., 2011). IP5 is closely located in the binding pocket of JA-Ile and co-ordinates three arginine residues of COI1 and R206 of the Jas peptide (Sheard et al., 2010). The IP5-free receptor complex is inactive. Previous pull-down experiments revealed that (+)-7-iso-JA-Ile is the most bioactive ligand (Fonseca et al., 2009). This is now substantiated by the crystal structure: most of the ligand is surrounded by COI1 residues, but the keto-group of JA in JA-Ile and the COOH-group of Ile can interact with the Jas domain (Sheard et al., 2010). Initial binding assays with labelled JA-Ile and COI1 protein showed a strong (50-fold) increase in binding and in specificity, if JAZ1 or JAZ6 were used as co-receptor complex component (Katsir et al., 2008b). Site-directed mutagenesis revealed essential amino acid residues for binding of the ligand in the binding pocket established by the COI1– JAZ interaction (Melotto et al., 2008).

Several defense genes are induced by JA. MeJA treatment induced eight defense genes consisting of PR1, PR2 (β-1,3-glucanase), PR3 (cuitinase), PR4 (wheatwin), PR5 (thaumatin-like protein), TaPERO (peroxidase), PR10, and TaGLP2a (germin-like) in wheat (Desmond et al., 2006). JA induces β-1,3-glucanase and thaumatin-like proteins in wheat (Jayaraj et al., 2004). JA/ET signaling pathway systems activate the transcription of several defense-related PR genes including PR1, PR2, ChiB (PR3-basic chitin), PR4 and germin-like proteins (Pre et al., 2008). JA signaling is involved in triggering the phytoalexin camalexin biosynthesis in Arabidopsis (Berr et al., 2010).

The Physiological Roles of JA in Plant Defense

For over 90 years, scientists and naturalists have observed that when plants survive pathogen infection they develop an increased resistance to subsequent infections (John Ryals et al., 1994). Plants can activate separate defense pathways depending on the type of pathogen encountered. The production of phytoalexins and other antimicrobial compounds can directly interfere with the survival and spread of the pathogen (Osbourn, 1996). Two inducible defense systems have been well-characterized in vascular plants: a localized hypersensitive response (HR) and plant-wide systemic acquired resistance (SAR) (Heath, 2000; Conrath, 2006. HR is characterized by ion fluxes, the generation of reactive oxygen species, and localized programmed cell death, which are governed by interactions among the products of pathogen avirulence genes and those of plant resistance genes (Heath, 2000). In SAR, a plant detects the presence of a pathogen and transmits a signal throughout the plant via the phloem, inducing changes in the expression of various pathogenesis-related (PR) genes (Conrath, 2006; van Loon et al., 2006). Once SAR is established, the plant is capable of mounting rapid responses to subsequent attacks from a wide range of pathogens and confers long-lasting protection against a broad spectrum of microorganisms.

The three phytohormones jasmonic acid (JA), salicylic acid (SA), and ethylene (ET) are known to play major roles in regulating plant defense responses against various pathogens, pests and abiotic stresses such as wounding by activating pathogenesis-related(PR)defense genes and establishing systemic resistance (Loake and Grant, 2007; Balbi and Devoto, 2008; Van Loon et al., 2006). Specifically, pathogenesis-related (PR) genes are shown to be associated with signaling pathways and subsequent establishment of systemic resistance. These PR genes encode various antimicrobial products, including β-1,3 glucanases and chitinases (Van Loon et al., 2006). SA plays a crucial role in plant defense and is generally involved in the activation of defense responses against bio-trophic and hemi-biotrophic pathogens as well as the establishment of systemic acquired resistance (SAR, Grant and Lamb, 2006). SA levels increase in pathogen-challenged tissues of plants and exogenous applications result in induction of pathogenesis related (PR) genes and enhanced resistance to a broad range of pathogens. By contrast, JA and ET are usually associated with defense against necotropic pathogens and herbivores insects. Although, SA and JA/ET defense pathways are mutually antagonistic, evidences of synergistic interactions have also been reported (Schenk et al., 2000; Kunkel and Brooks, 2002; Beckers and Spoel, 2006; Mur et al., 2006). This suggests that the defense signaling network activated and utilized by the plant is dependent on the nature of the pathogen and its mode of pathogenicity.

Jasmonic Acid and Systemic Induced Resistance

JA is an essential signal for plant defense against insect herbivory and it provides a major mechanism of induced defenses against insect's herbivores and a wide spectrum of pathogen species, especially necrotrophic fungi. The defense property of jasmonates to various insect herbivores has been extensively studied in the past decades (Browse, 2009; Howe and Jander, 2008). The major lines of evidence that points the significance of JA as a defense hormone in plants are:

- Mechanical wounding or damage caused by herbivore feeding results in rapid accumulation of JAs at the site of wounding (Glauser et al., 2008). Successive feeding on the leaves causes steady increase of JA content throughout the entire plant (Reymond et al., 2004).
- Wounding or exogenous application of JA/MeJA generally up-regulates the genes involved in JA biosynthesis (Mueller, 1997; Leon and Sanchez-Serrano, 1999). Most JA biosynthesis genes such as *LOX2*, *LOX3*, *LOX4*, *AOS*, *OPR3*, and *AOC3* and signaling genes such as *MYC2*, *JAZ1*, *JAZ2*, *JAZ5*, *JAZ6*, *JAZ7*, *JAZ8*, *JAZ9,* and *JAZ10* were found highly inducible in response to wounding, MeJA treatment, and herbivore feeding (Chung et al., 2008).
- Insect feeding or wounding induces hundreds of defense-related genes in JA dependent manner, including genes involved in pathogenesis, indole glucosinolate metabolism, and detoxification (Reymond et al., 2004).
- Insect feeding, wounding, or MeJA treatment activates synthesis of anti-insect substance, e.g., proteinase inhibitors (PIs) in *Arabidopsis* (Farmer et al., 1992), nicotine in tobacco, papain inhibitor(s) in tomato (Bolter, 1993), vinblastine in rose periwinkle (*Catharanthus roseus*), artemisinin in annual wormwood (*Artemisia annua*) (De Geyter et al., 2012), and poisonous secondary metabolites such as glucosinates and camalexin in *Arabidopsis*.
- JA biosynthesis or perception mutants of *Arabidopsis* such as, *fad3-2 fad7-2 fad8*, *aos*, *opr3*, *jar1,* and *coi1,* as well as those from other species such as tomato *jar1*, and maize *opr7 opr8* are highly susceptible to insect attack (McConn et al., 1997; Laudert and Weiler, 1998; Stintzi et al., 2001; Staswick et al., 1998; Xie et al., 1998; Li et al., 2004; Yan et al., 2012). These JA mutants are shown to be compromised in resistance to a wide range of arthropod herbivores including caterpillars (*Lepidoptera*), beetles (*Coleoptera*), thrips (*Thysanoptera*), leafhoppers (*Homoptera*), spider mites (*Acari*), fungal gnats (*Diptera*), and mirid bugs (*Heteroptera*) (Howe and Jander, 2008).

- Exogenous application of JA or MeJA can elevate resistance of a number of plant species to insects attack (Avdiushko et al., 1997). The JA precursor OPDA also contributes to plant defense against insect attacks (Stintzi et al., 2001).
- When attacked by herbivores, plants can rapidly release volatile organic compounds (VOC, consisting mainly of fatty acid derived products and terpenes) and green leafy volatiles (GLV, including mainly of (*Z*)-*3*- hexenal, (*Z*)-3-hexenol, and (*Z*)-3-hexenyl acetate) (Figure 7). These can effectively induce direct defense response-activation of JA biosynthesis pathway in the attacked and neighboring plants, and indirect defense response-attraction of insect enemies that parasitize or prey on feeding insects (Paré and Tumlinson, 1999; Engelberth et al., 2004).

The importance of cell wall fragments in elicitation was supported by the finding that cellulysin, a mixture of several cell wall-degrading enzymes from the plant parasitic fungus *Trichoderma viride*, can induce several JA-responsive volatiles in lima bean (Phaseolus lunatus) (Piel et al., 1997). The action of cellulysin is followed by a rapid increase in endogenous JA (Koch et al., 1999) (Figure 8).

Wounding by herbivores leads to (1) direct defence via synthesis of toxic compounds as well as toxic or anti-nutritional proteins such as proteinase inhibitors (PIs) and (2) indirect defence by release of volatiles to affect the attraction of carnivores, parasitoids and predators or by altered oviposition of herbivores (Figure 8) (Wasternack and Hause, 2002; Howe and Jander, 2008). In this event plants become immunized by the PI expression that affects protein digestion in the herbivore gut. Of the herbivore and JA-induced proteins, threonine deaminase 1 (TD1) is required for the formation of isoleucine. An interesting adaptive evolution by gene duplication of TD has been demonstrated for tomato (Chen et al., 2007; Gonzales-Vigil et al., 2011). The proteolytic cleavage of the regulatory domain of TD2 may be attributed to plant resistance in response to herbivory.

Local wounding by herbivores leads to a burst in newly synthesized JA. The constitutive occurrence of enzyme proteins involved in JA biosynthesis in all leaf tissues (as in Arabidopsis; Stenzel et al., 2003b) or in vascular bundles (as in tomato; Hause et al., 2000, 2003) may be ascribed to such an immediate rise in JA levels within several minutes (Glauser et al., 2008; Mielke et al., 2011), whereas later (from 15 min onwards) the transcriptional machineries for the expression of LOX, AOS, AOC, OPR3 and JAZs are activated (Koo and Howe, 2009). In this signaling network compounds of insects oral secretions, such as volicitin (Bonaventure et al., 2011b), the peptide systemin, H_2O_2, NO and ET, act as additional positive signals (Koo and Howe, 2009).

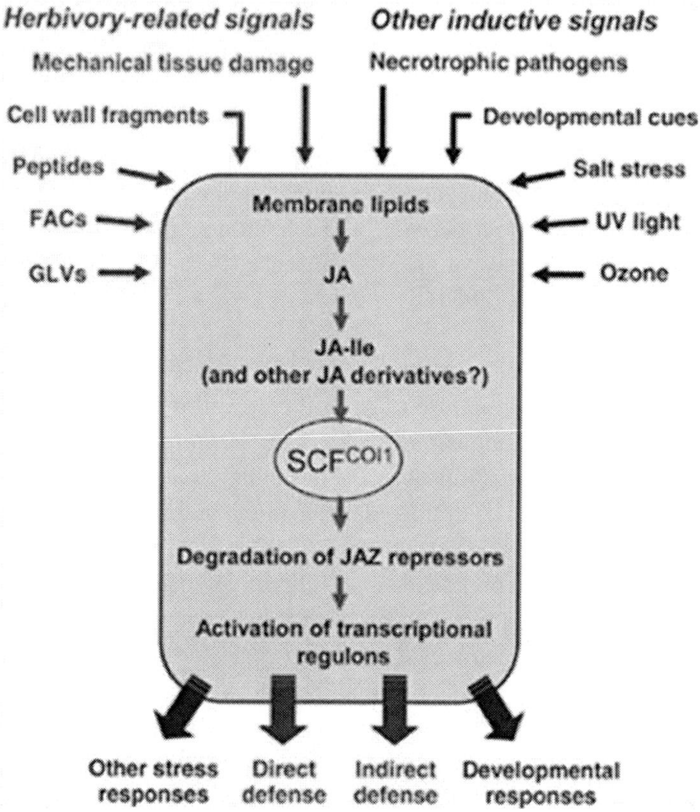

Source: Howe and Jander, 2008.

Figure 8. JA regulates numerous physiological processes in response to environmental and developmental cues. FACs, fatty acid-amino acid conjugates; GLVs, green leafy volatiles.

Systemin was the first peptide identified as a signaling compound in wounded tomato leaves (Pearce et al., 1991). It is an 18-amino acid polypeptide that is released upon wounding from a 200-amino acid precursor ('prosystemin') and that leads to the release of linolenic acid. This activates the octadecanoid signaling cascade (Ryan, 2000). Both JA (Zhang and Baldwin, 1997) and systemin (Ryan, 2000) can be transported in the phloem and thus might act as systemic signals. Initially, systemin was thought to be a systemic signal involved in long distance signaling. After two decades of research on wound-induced systemic response, systemin is thought to have a minor role by amplification of systemic wound signaling in a tissue-specific manner (Hind et al., 2010; Sun et al., 2011a). Due to the central role of JA/JA-Ile in SCF^{COI1}-mediated signaling, one amplification activity of systemin seems to be its positive regulation of the expression of AOC and JA formation upon wounding (Stenzel et al., 2003a). This activation of JA biosynthesis by systemin involves activity of MPK1, MPK2 and MPK3 (Kandoth et al., 2007).

The joint role of (pro) systemin and the octadecanoid pathway in the generation of a phloem-mobile signal for systemic induction of defense genes is consistent with their localization in the vasculature of tomato plants. Several of the octadecanoid pathway enzymes – LOX, AOS, and AOC – are located in the companion cell sieve element complex (Hause et al., 2000, 2003a), whereas prosystemin accumulates in cells of the phloem parenchyma (Jacinto et al., 1997). The presence of prosystemin and octadecanoid pathway enzymes in different cell types of the vascular bundle suggests a model in which systemin is released from phloem parenchyma cells in response to wounding, is then perceived at the cell surface of neighboring companion cells, where it triggers the activation of the octadecanoid pathway for the production of the systemic signal, and its release into the sieve elements for long-distance transport (Schilmiller and Howe, 2005; Wasternack, 2006). Following the production of JA at the site of tissue damage and transmission of the long distance signal, the activation of defense responses in systemic tissues requires COI1 (Li et al., 2002, 2003).

Mitogen-activated protein kinases (MAPKs) and ca2+-binding sensory proteins concomitantly and independently play important roles in mediating herbivory responses. Herbivory- or wounding-related MAPKs, SA-induced protein kinase (SIPK) and wound-induced protein kinase (WIPK) were the first MAPKs identified in tobacco (Seo, S et al., 1995). Transcripts of the WIPK gene begin to accumulate one minute after mechanical wounding, leading to the induced production of JA-inducible gene transcripts (Seo, s et al., 1999). In addition, SIPK is also known to be involved in both JA and ethylene production (Kim et al., 2003), whereas activation of SIPK after wounding is associated with increased tyrosine phosphorylation but not with increases in SIPK mRNA or protein levels. Kandoth et al. reported that co-silencing of MPK1 and MPK2 (orthologues of WIPK and SIPK genes) in tomato over expressing prosystemin weakens proteinase inhibitor-associated defense against the specialist herbivore Manduca sexta (Kandoth et al., 2007).

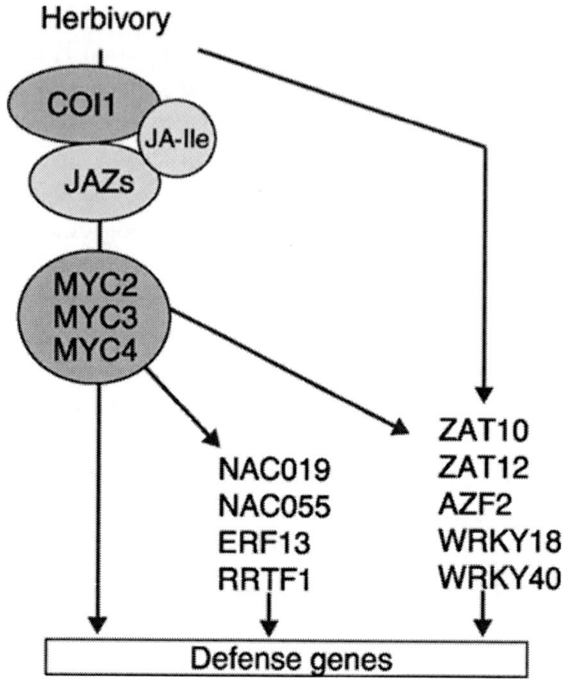

Source: Fabian Schweizer et al., 2013.

Figure 9. A model for the transcriptional network in defense against insects. In response to herbivory, plants produce JA-Ile. This hormone is detected by its receptor COI1 that in turn degrades JAZ repressors (not shown) to allow the transcriptional activity of MYC2, MYC3, and MYC4 TFs. As a consequence, MYCs activate the expression defense genes and downstream TFs. In addition, herbivory induces the expression of several TFs, which partially depend on COI1 and MYCs. How these TFs are regulated and which are their target genes remains unknown. This model only contains TFs described in this study.

During insect herbivory, plants induce about 1000 genes, of which roughly 65% are regulated by the JA-pathway (Halitschke et al., 2001; Reymond et al., 2004; de Vos et al., 2005). Following JA-Ile perception by COI1, repression of MYC2, MYC3, and MYC4 by JAZs is released allowing the transcription of defense genes. TFs activate defense gene expression in JA-dependent and JA-independent manner (Figure 9). For the JA-dependent pathway, MYC2/MYC3/MYC4 plays a quantitatively important role by directly activating defense genes or by activating downstream TFs. In parallel, a JA-independent pathway triggers WRKYs and Zinc-finger TFs expression to provide additional defense. These findingssuggests only a fraction of all TFs involved in defense against herbivory.

JA and Induced Systemic Resistance (ISR) Against Microbial Pathogens

JA is an essential phytohormone for defense response against a wide spectrum of pathogens, alone or in combination with other hormones, such as ET, SA, and ABA (Browse, 2009; Adie et al., 2007). Although all plant hormones including GA, auxin (IAA), and brassinosteroids (BR) may be involved in plant defense responses against pathogens (Smith et al., 2009), numerous studies have shown that SA, JA, and ET are the major players in induced resistance of plants (Dong, 1998; Kunkel and Brooks, 2002). Rhizobacteria-mediated induced systemic resistance (ISR) in plants primarily depends on JA and ET (Pieterse et al., 1998). Regarding the relationship of JA with ET, the widely held belief is that ET acts synergistically with JA in the activation of responses to pathogens (Lorenzo and Solano, 2005). Several defense-related genes including *PR1*, *PR3*, *PR4*, *PR5,* and *PDF1.2* are synergistically induced by JA and ET (Lorenzo et al., 2003).

Considering only the role of JA, it has been frequently demonstrated to be an indispensible phytohormone signal for resistance/susceptibility to several diseases caused by fungal, bacterial, and viral pathogens. JA mediated resistance to necrotrophic pathogens such as *Botrytis cinerea* and *Alternaria brassicicola* has been reported by Thomma et al., (1998). *Arabidopsis* fatty acid desaturase triple mutant *fad3-2 fad7-2 fad8*, deficient in biosynthesis of the JA precursor linolenic acid, is more susceptible to the root pathogen *Pythium mastophorum*; 90% of the triple mutant plants did not survive the infection as compared to only 10% of wild-type plants (Vijayan et al., 1998). Exogenously applied MeJA reduced death rate of *fad3-2 fad7-2 fad8*. Another *Arabidopsis* JA mutant *jar1* (*jasmonic-acid resistant 1*), shows reduced sensitivity to jasmonates and deficient JA signaling (Staswick et al., 1998). Both *fad3-2 fad7-2 fad8* and *jar1* plants exhibit enhanced susceptibility to *Pythium irregulare* (Staswick et al., 1998). In maize, the double mutant *opr7 opr8*, deficient in JA biosynthesis, showed extreme susceptibility to *Pythium aristosporium* (Yan et al., 2012).

JAs may have positive roles in plant resistance against viruses. Members of the geminivirus family are plant viruses with circular, single-stranded DNA genomes that infect a wide range of plant species and cause extensive yield losses in important crops such as tomato, maize, and cotton. In *Arabidopsis*, exogenous application of jasmonates reduces susceptibility to geminivirus infection (Lozano-Durán et al., 2011). In a case of turnip crinkle virus (TCV), SA, but not JA/ET, is required for the development of hypersensitive reaction

(HR) and systemic resistance in *Arabidopsis* (Kachroo et al., 2000). However, applying 60 μM JA and then 100 μM SA 24 h later, enhanced resistance to *Cucumber mosaic virus* (CMV), *Tobacco mosaic virus* (TMV), and TCV in *Arabidopsis*, tobacco, tomato, and hot pepper (Shang et al., 2011), indicating JA and SA have additive positive effects to resistance against RNA viruses.

JA Activate Secondary Metabolism Beneficial to Development and Defense

Secondary metabolites play diverse roles in plants. For example, flowers synthesize and accumulate anthocyanin pigments in petals to attract pollinating insects. In addition, anthocyanins absorb visible as well as UV radiation and are effective antioxidants and scavengers of reactive oxygen species, protecting plant tissues from the effects of excess incidental visible or UV-B radiation and oxidative stress (Quina et al., 2009). Other secondary metabolites such as polyamines, quinones, terpenoids, alkaloids, phenylpropanoids, and glucosinolates act as phytoalexins to protect plants against microorganisms or herbivores (Chen et al., 2006). Gaseous MeJA enhance production of anthocyanins in soybean seedlings (Franceschi and Grimes, 1991). Wounding or MeJA treatment activates rapidly the expression of anthocyanin biosynthesis genes and increase anthocyanin level in the detached corolla of *Petunia hybrida* (Moalem-Beno et al., 1997). In *Arabidopsis*, JA or MeJA treatment strongly enhances anthocyanin accumulation in the shoots, especially in the petiole of the seedling (Lorenzo et al., 2004) and this JAs-activating anthocyanin accumulation depends on COI1- mediated JA signaling pathway (Shan et al., 2009). JA induces anthocyanin biosynthesis via up-regulation of the 'late' anthocyanin biosynthetic genes *DFR*, *LDOX*, and *UF3GT* (Shan et al., 2009). JA coincidently activates anthocyanin biosynthetic regulators such as transcription factors *PAP1*, *PAP2*, and *GL3*. Either these biosynthetic genes or transcription factors are *COI1*-dependent (Shan et al., 2009).

JAs also effectively activate defensive metabolites against insects or pathogens. MeJA application strongly induced anti-insect protein accumulation such as proteinase inhibitors I and II (PI-I, II) (Farmer et al., 1992) and vegetative storage protein (VSP) (Liu et al., 2005). Nicotine, an alkaloid toxic to most insects by interfering with the transmitter substance between nerves and muscles, widely exists in tobacco (*Nicotiana tabacum*) and related species. Exogenous application of JA or wounding of leaves activate nicotine biosynthesis in a *COI1*- and *MYC2*-dependent manner, indicating JA signal is required in tobacco to control nicotine metabolism (Shoji et al., 2008; Shoji and Hashimoto, 2011). In *Arabidopsis*, camalexin (3-thiazol-2'yl-indole) is the main phytoalexin induced by a variety of microorganisms including bacteria, fungi, and oomycetes. JA signaling is required for the activation of camalexin synthesis in response to infection by *P. syringae* pv. *maculicola* ES4326 (Zhou et al., 1999).

Glucosinolates are a group of thioglucosides found in all cruciferous plants such as *Arabidopsis* and *Brassica napus*. The hydrolysed products of glucosinolates contribute to plant defense against microorganisms. MeJA treatment increases total glucosinolate content in leaves of *B. napus* up to 20 fold (Doughty et al., 1995). In *Arabidopsis*, accumulation of camalexin and indole glucosinolates can be trigged by elicitors from the plant pathogen *Erwinia carotovora,* and this induction effect is *COI1*- dependent (Brader et al., 2001). There are also a number of examples that jasmonates effectively activate secondary metabolites in medicinal plant species such as artemisinin synthesis in *Artemisia annua* and vinblastine (an alkaloid) in *Catharanthus roseus* (De Geyter et al., 2012). All pathways of the above metabolites, including nicotine, camalexin, glucosinolates, artemisinin, and vinblastine belong to JA-elicited plant secondary metabolism, which is regulated by JA-signaling components such as *COI1*, *MYC2*, *ERF1* and *JAZs* (De Geyter et al., 2012).

CONCLUSION

JA biosynthesis enzymes showed 'self-activation', in which the final product, JA, positively regulate the enzyme activity of this pathway. Currently, only (+) -7-iso-jasmonoyl-L-Ile has been conclusively shown to function as the bioactive ligand to JA signaling machinery SCF^{COI1}/JAZs complex. The molecular mechanism of JA signal perception and transduction was found to mimic many aspects of the auxin signaling process. In the presence of low levels of JA, JAZ proteins repress the expression of JA-responsive genes by interacting directly with the bHLH (basic helix-loophelix) transcription factors MYC2, MYC3, and MYC4, which are positive regulators of JA responses. When JA levels increased, the bioactive form of ligand JA-Ile promotes binding of JAZs to SCF^{COI1} to form SCF^{COI1}-JAZ-JA-Ile reception complex and subsequent degradation of JAZ repressors via the ubiquitin/26S proteasome pathway, resulting in de-repression of primary response genes. A number of recent studies found a wide spectrum of JA functions in plant including the regulation of developmental and defense processes, such as, resistance against insects and pathogens, root growth, fruit/seed maturation, leaf senescence, anthocyanin pigmentation, sex determination (of monoecious plant), female or reproductive organ formation. Plant signaling pathways driven by Jasmonic acid, Salicylic acid, Ethylene, Abscisic acid, Gibberllic acid, reactive oxygen species (ROS), and Nitric oxide induce changes in the expression of defense genes, which lead to metabolic changes enhancing plant defenses in response to attack by aphids. Furthermore, the activation of transcription factors often enables a cross-talk between antagonistic and synergistic relationships between the pathways, and defines gene expression profiles of induced resistance. Jasmonic acid and its derivatives have been shown to induce resistance to various insects including caterpillars, beetles, thrips, leafhoppers, spider mites and mired bugs in rice, cotton, wheat, tobacco, tomato and soybean. Simultaneous activation of ISR and SAR by JA, ET and SA results in an enhanced level of induced protection in plants against various pathogens.

REFERENCES

Acosta, I.F., Laparra, H., Romero, S.P., Schmelz, E., Hamberg, M., Mottinger, J.P., Moreno, M.A., and Dellaporta, S.L.2009. Tasselseed1 is a lipoxygenase affecting jasmonic acid signaling in sex determination of maize. *Science* 323:262-265.

Adie, B.A., Perez-Perez, J., Perez-Perez, M.M., Godoy, M., Sanchez-Serrano, J.J., Schmelz, E.A., and Solano, R. 2007. ABA is an essential signal for plant resistance to pathogens affecting JA biosynthesis and the activation of defenses in Arabidopsis. *Plant Cell* 19:1665-1681.

Agrawal, G.K., Tamogami, S., Han, O., Iwahashi, H., and Rakwal, R. 2004. Rice octadecanoid pathway. *Biochem Biophys Res Commun*. 317:1-15.

Anderson, J.P., Badruzsaufari, E., Schenk, P.M., Manners, J.M., Desmond, O.J., Ehlert, C., Maclean, D.J., Ebert, P.R., and Kazan, K. 2004. Antagonistic interaction between abscisic acid and jasmonate-ethylene signaling pathways modulates defense gene expression and disease resistance in Arabidopsis. *Plant Cell* 16:3460-3479.

Anderson, M.X., Hamberg, M., Kourtchenko, O., Brunnstrom, A., McPhail, K.L., Gerwick, W.H., Gobel, C., Feussner, I., and Ellerstrom, M., 2006. Oxylipin profiling of the hypersensitive response in *Arabidopsis thaliana*.

Formation of a novel oxo-phytodienoic acid containing galactolipid, arabidopsides. *E. J. Biol. Chem.* 281: 31528-31537.

Andreou, A., Brodhun, F., and Feussner, I. 2009. Biosynthesis of oxylipins in non-mammals. *Prog. Lipid Res.* 48:148-170.

Andresen, I., Becker, W., Schliiter, K., Burges, J., Parthier, B., Apel, K. 1992. The identification of leaf thionin as one of the main jasmonate-induced proteins of barley (*Hordeum vulgare*). *Plant Mo1. Biol.* 19: 193-204.

Arimura, G., Ozawa, R., Shimoda, T., Nishioka, T., Boland, W., Takabayashi, J. 2000. Herbivory-induced volatiles elicit defence genes in lima bean leaves. *Nature* 406: 512–515.

Avanci, N.C., Luche, D.D., Goldman, G.H., and Goldman, M.H. 2010. Jasmonates are phytohormones with multiple functions, including plant defense and reproduction. *Genet. Mol. Res.* 9:484-505.

Avdiushko, S.A., Brown, G.C., Dahlman, D.L., and Hildebrand, D.F. 1997. Methyl jasmonate exposure induces insect resistance in cabbage and tobacco. *Environmental Entomology* 26: 642-654.

Balbi, V, Devoto, A. 2008. Jasmonate signalling network in *Arabidopsis thaliana*: crucial regulatory nodes and new physiological scenarios. *New Phytol.* 177: 301–318.

Bannenberg, G., Martinez, M., Hamberg, M., and Castresana, C. 2009. Diversity of the enzymatic activity in the lipoxygenase gene family of *Arabidopsis thaliana*. *Lipids* 44:85-95.

Beckers, G.J., and Spoel, S.H. 2006. Fine-tuning plant defence signalling: salicylate versus jasmonate. *Plant. Biol.* (Stuttg) 8:1-10.

Bell, E., Creelman, R.A., and Mullet, J.E. 1995. A chloroplast lipoxygenase is required for wound-induced jasmonic acid accumulation in Arabidopsis. *Proc. Natl. Acad. Sci. USA* 92:8675-8679.

Bell, E., Mullet, J.E. 1991. Lipoxygenase gene expression is modulated in plants by water deficit, wounding, and methyl jasmonate. *Mol. Gen. Genet.* 230: 456-462.

Berr, A. et al. 2010. Arabidopsis histone methyltransferase SET DOMAIN GROUP8 mediates induction of the jasmonate/ethylene pathway genes in plant defense response to necrotrophic fungi. *Plant. Physiol.* 154: 1403-1414.

Blechert, S., Bockelmann, C., F¨ulein, M., Von Schrader, T., Stelmach, B., Niesel, U., Weiler, E.W. 1999. Structure-activity analyses reveal the existence of two separate groups of active octadecanoid in elicitation of the tendril-coiling response of *Bryonia dioica* Jacq. *Planta* 207:470–479.

Bolter, C.J. 1993. Methyl Jasmonate Induces Papain Inhibitor(s) in Tomato Leaves. *Plant. Physiol.* 103: 1347-1353.

Bonaventure, G., Gfeller, A., Proebsting, W.M., Hortensteiner, S., Chetelat, A., Martinoia, E., and Farmer, E.E., 2007. A gain-of-function allele of TPC1 activates oxylipin biogenesis after leaf wounding in Arabidopsis. *Plant. J.* 49:889-898.

Bonaventure, G., Van Doorn, A., Baldwin, I.T. 2011b. Herbivore-associated elicitors: FAC signaling and metabolism. *Trends in Plant Science* 16: 294–299.

Brader, G., Tas, E., Palva, E.T. 2001. Jasmonate-dependent induction of indole glucosinolates in Arabidopsis by culture filtrates of the nonspecific pathogen *Erwinia carotovora. Plant Physiology* 126: 849-860.

Breithaupt, C., Kurzbauer, R., Lilie, H., Schaller, A., Strassner, J., Huber, R., Macheroux, P., Clausen, T., 2006. Crystal structure of 12-oxophytodienoate reductase 3 from tomato: self-inhibition by dimerization. *Proc. Natl. Acad. Sci. USA* 103: 14337– 14342.

Breithaupt, C., Kurzbauer, R., Schaller, F., Stintzi, A., Schaller, A., Huber, R., Macheroux, P., and Clausen, T. 2009. Structural basis of substrate specificity of plant 12-oxophytodienoate reductases. *J. Mol. Biol.* 392:1266-1277.

Breithaupt, C., Strassner, J., Breitinger, U., Huber, R., Macheroux, P., Schaller, A., Clausen, T., 2001. X-ray structure of 12-oxophytodienoate reductase 1 provides structural insight into substrate binding and specificity within the family of OYE. *Structure* 9: 419–429.

Browse, J. 2009. Jasmonate passes muster: a receptor and targets for the defense hormone. *Annu Rev Plant Biol.* 60:183-205.

Browse, J. 2009. The power of mutants for investigating jasmonate biosynthesis and signaling. *Phytochemistry* 70:1539-1546.

Buer, C.S., Imin, N., Djordjevic, M.A. 2010. Flavonoids: New roles for old molecules. *J. Integr. Plant. Biol.* 52: 98–111.

Buseman, C.M., Tamura, P., Sparks, A.A., Baughman, E.J., Maatta, S., Zhao, J., Roth, M.R., Esch, S.W., Shah, J., Williams, T.D., et al. 2006. Wounding stimulates the accumulation of glycerolipids containing oxophytodienoic acid and dinor-oxophytodienoic acid in *Arabidopsis* leaves. *Plant. Physiol.* 142:28–39.

Chen, H., Gonzales-Vigil, E., Wilkerson, C.G., Howe, G.A. 2007. Stability of plant defense proteins in the gut of insect herbivores. *Plant. Physiology* 143: 1954–1967.

Chen, H., Jones, A.D., and Howe, G.A. 2006. Constitutive activation of the jasmonate signaling pathway enhances the production of secondary metabolites in tomato. *FEBS Lett.* 580: 2540-2546.

Cheong, J.J., and Choi, Y.D., 2003. Methyl jasmonate as a vital substance in plants. *Trends in genetics* 19:409-413.

Chini, A., Fonseca, S., Fernandez, G., Adie, B., Chico, J.M,, Lorenzo, O., Garcia-Casado, G., Lopez- Vidriero, I., Lozano, F.M., Ponce, M.R., Micol, J.L., and Solano, R. 2007. The JAZ family of repressors is the missing link in jasmonate signalling. *Nature* 448:666-671.

Choi, D., Ward, B.L., Bostock, R.M. 1992. Differential induction and suppression of potato 3-hydroxy-3-methylglutaryl coenzyme A reductase genes in response to *Phytophthoru infestam* and to its elicitor arachidonic acid. *Plant Cell* 4: 1333-1344.

Chung, H. S., Koo, A.J.K., Gao, X., Jayanty, S., Thines, B., Jones, A. D., and Howe, G. A. 2008. Regulation and function of *Arabidopsis* JASMONATE ZIM-domain genes in response to wounding and herbivory. *Plant Physiol.*146: 952–964.

Chung, H.S., Howe, G.A. 2009. A critical role for the TIFY motif in repression of jasmonate signaling by a stabilized splice variant of the JASMONATE ZIM-domain protein JAZ10 in Arabidopsis. *The Plant Cell* 21: 131–145.

Conrath, U., 2006. Systemic acquired resistance. *Plant Signaling and Behavior* 1: 179–184.

Creelman, R.A., and Mullet, J.E., 1995. Jasmonic acid distribution and action in plants: regulation during development and response to biotic and abiotic stress. *Proc. Natl. Acad. Sci.* USA 92:4114-4119.

Creelman, R.A., and Mullet, J.E., 1997. Biosynthesis and Action of Jasmonates in Plants. *Annu Rev Plant Physiol Plant Mol Biol.* 48:355-381.

Creelman, R.A., Tierney, M.L., Mullet, J. 1992. Jasmonic acid/methyl jasmonate accumulate in wounded soybean hypocotyls and modulate wound gene expression. *Proc. Natl. Acad. Sci. USA* 89:4938–4941

Cruz Castillo, M., Martinez, C., Buchala, A., Metraux, J.P., Leon, J. 2004. Gene-specific involvement of beta-oxidation in wound-activated responses in *Arabidopsis*. *Plant. Physiol.* 135:85–94.

Czapski, J., Saniewski, M. 1992. Stimulation of ethylene production and ethylene-forming enzyme activity in fruits of the non-ripening or and in tomato mutants by methyl jasmonate. *J. Plant. Physiol.* 139: 265-268.

Dathe, W., Ronsch, H., Preiss, A., Schade, W., Sembdner, G., Schreiber, K. 1981. Endogenous plant hormones of the broad bean, *Vicia faba* L. (–)-Jasmonic acid, a plant growth inhibitor in pericarp. *Planta* 153:530–535.

De Geyter, N., Gholami, A., Goormachtig, S., and Goossens, A. 2012. Transcriptional machineries in jasmonate-elicited plant secondary metabolism. *Trends Plant. Sci.* 17: 349- 359.

De Vos, M., Van Zaanen, W., Koornneef, A., Korzelius, J.P., Dicke, M., Van Loon, L.C., Pieterse, C.M.J. 2006. Herbivore-induced resistance against microbial pathogens in Arabidopsis. Plant. Physiol. 142 352–363

Delker, C., Stenzel, I., Hause, B., Miersch, O., Feussner, I., and Wasternack, C. 2006. Jasmonate biosynthesis in *Arabidopsis thaliana* enzymes, products, regulation. *Plant. Biol.* (Stuttg) 8:297-306.

Delker, C., Zolman, B.K., Miersch, O., and Wasternack, C. 2007. Jasmonate biosynthesis in Arabidopsis thaliana requires peroxisomal beta-oxidation enzymes--additional proof by properties of pex6 and aim1. *Phytochemistry* 68:1642-1650.

Demole, E., Lederer, E., and Mercier, D. 1962. Isolement et détermination de la structure du jasmonate de méthyle, constituant odorant charactéristique de l'essence de jasmin. *Helvetica Chimica Acta* 45: 675-685.

Desmond, O.J., C.I. Edgar, J.M. Manners, D.J. Maclean, P.M. Schenk, and K. Kazan. 2006. Methyl jasmonate induced gene expression in wheat delays symptom development by the crown rot pathogen *Fusarium pseudograminearum. Physiological and Molecular Plant Pathology* 67: 171-179.

Dodd, A.N., Kudla, J., Sanders, D. 2010. The language of calcium signaling. *Annu. Rev. Plant. Biol.* 61: 593–620.

Dong, X. 1998. SA, JA, ethylene, and disease resistance in plants. *Curr Opin Plant. Biol.* 1: 316- 323.

Doughty, K.J., Kiddle, G.A., Pye, B.J., Wallsgrove, R.M., and Pickett, J.A. 1995. Selective induction of glucosinolates in oilseed rape leaves by methyl jasmonate. *Phytochem.* 38:347-350.

Dudareva, N., D'Auria, J.C., Nam, K.H., Raguso, R.A., Pichersky, E. 1998. Acetyl-CoA: benzyl alcohol acetyltransferase: an enzyme involved in floral scent production in *Clarkia breweri.* Plant J. 14: 297–304.

Ellinger, D., Stingl, N., Kubigsteltig, Bals, T., Juenger, M., Pollmann, S., Berger, S., Schuenemann, D., and Mueller, M.J. 2010. DONGLE and DEFECTIVE IN ANTHER DEHISCENCE1 lipases are not essential for wound-and pathogen-induced jasmonate biosynthesis: redundant lipases contribute to jasmonate formation. *Plant. Physiol.* 153:114.

Ellis, C., and Turner, J.G., 2001. The Arabidopsis mutant cev1 has constitutively active jasmonate and ethylene signal pathways and enhanced resistance to pathogens. *Plant. Cell* 13:1025-1033.

Engelberth, J., Alborn, H.T., Schmelz, E.A., and Tumlinson, J.H. 2004. Airborne signals prime plants against insect herbivore attack. *Proc. Natl. Acad. Sci.* USA. 101:1781-1785.

Fabian Schweizer., Natacha Bodenhausen., Steve Lassueur., Frédéric, G., Masclaux and Philippe Reymond. 2013. Differential contribution of transcription factors to *Arabidopsis thaliana* defense against *Spodoptera littoralis. Front. Plant Sci.* 4:1-12.

Farmaki, T., Sanmartin, M., Jimenez, P., Paneque, M., Sanz, C., Vancanneyt, G., Leon, J., Sanchez-Serrano, J.J. 2007. Differential distribution of the lipoxygenase pathway enzymes within potato chloroplasts. *J. Exp. Bot.* 58:555–568.

Farmer, E.E., Johnson, R.R., and Ryan, C.A. 1992. Regulation of expression of proteinase inhibitor genes by methyl jasmonate and jasmonic Acid. *Plant Physiol.* 98: 995-1002.

Farmer, E.E. 2007. Plant biology: jasmonate perception machines. *Nature* 448:659–660.

Faure, D., Vereecke, D., Leveau, J.H.J. 2009. Molecular communication in the rhizosphere. *Plant. Soil* 321: 279–303.

Feussner, I., Wasternack, C. 2002. The lipoxygenase pathway. *Annu. Rev. Plant. Biol.* 53:275–297.

Feys, B.J.F., Benedetti, C.E., Penfold, C.N., Turner, J.G. 1994. *Arabidopsis* mutants selected for resistance to the phytotoxin coronatine are male sterile, insensitive to methyl jasmonate, and resistant to a bacterial pathogen. *Plant. Cell* 6:751–759.

Fonseca, S., Chini, A., Hamberg, M., Adie, B., Porzel, A., Kramell, R., Miersch, O., Wasternack, C., Solano, R., 2009. (+)-7-iso-Jasmonoyl-L-isoleucine is the endogenous bioactive jasmonate. *Nat. Chem .Biol.* 5: 344-350.

Franceschi, V.R., and Grimes, H.D. 1991. Induction of soybean vegetative storage proteins and anthocyanins by low-level atmospheric methyl jasmonate. *Proc. Natl. Acad. Sci.* USA 88:6745-6749.

Fulda, M., Schnurr, J., Abbadi, A., Heinz, E., Browse, J. 2004. Peroxisomal Acyl-CoA synthetase activity is essential for seedling development in *Arabidopsis thaliana*. *Plant. Cell* 16:394–405.

Garbe, L.A., Barbosa, de., Almeida, R., Nagel, R., Wackerbauer, K., and Tress, R. 2006. Dual positional and Stereospecificity of lipoxygenase isoenzymes from germinating barley (green malt): biotransformation of free and esterified linoleic acid. *J. Agric. Food. Chem.* 54:946-955.

Gfeller, A., Dubugnon, L., Liechti, R., and Farmer, E.E. 2010. Jasmonate Biochemical Pathway. Sci Signal 3:cm3.

Glauser, G., Grata, E., Dubugnon, L., Rudaz, S., Farmer, E.E,, and Wolfender, J.L. 2008. Spatial and temporal dynamics of jasmonate synthesis and accumulation in Arabidopsis in response to wounding. *J. Biol. Chem.* 283: 16400-16407.

Göbel, C., and Feussner, I. 2009. Methods for the analysis of oxylipins in plants. *Phytochemistry* 70:1485-1503.

Gonzales-Vigil, E., Bianchetti, C.M., Phillips, G.N., Howe, G.A. 2011. Adaptive evolution of threonine deaminase in plant defense against insect herbivores. *Proceedings of the National Academy of Sciences of the USA* 108: 5897–5902.

Grant, M., Lamb, C. 2006. *Systemic immunity. Curr. Opin. Plant. Biol* 9: 414–*420.*

Gundlach, H., Muller, M.J., Kutchan, T.M., Zenk, M.H. 1992. Jasmonic acid as a signal transducer in elicitor-induced plant cell cultures. *Proc. Natl. Acad. Sci.* USA 89:2389–2393.

Guranowski, A., Miersch, O., Staswick, P.E., Suza, W., Wasternack, C.*2007.* Substrate specificity and products of side-reactions catalyzed by jasmonate: amino acid synthetase (JAR1).*FEBS Letters* 581*:*815-820.

Ha, S.B., Lee, B.C., Lee, D.E., Kuk, Y.I., Lee, A.Y., Han, O., and Back, K. 2002. Molecular characterization of the gene encoding rice allene oxide synthase and its expression. *Biosci Biotechnol Biochem*. 66:2719-2722.

Halitschke, R., Ursula Schittko., and Ian, T., Baldwin, 2001. Molecular Interactions between the Specialist Herbivore *Manduca sexta* (Lepidoptera, Sphingidae) and Its Natural Host *Nicotiana attenuata*. III. Fatty Acid-Amino Acid Conjugates in Herbivore Oral Secretions Are Necessary and Sufficient for Herbivore-Specific Plant Responses *Plant. Physiol.* 125 (2): 711-717.

Hamberg, M., Fahlstadius, P., 1990. Allene oxide cyclase: a new enzyme in plant lipid metabolism. *Arch. Biochem. Biophys.* 276: 518–526.

Harms, K., Atzorn, R., Brash, A., Kuhn, H., Wasternack, C., Willmitzer, L., Pena-Cortes, H. 1995. Expression of a flax allene oxide synthase cDNA leads to increased endogenous jasmonic acid (JA) levels in transgenic potato plants but not to a corresponding activation of JA-responding genes. *Plant Cell* 7:1645-1654.

Hause, B., Hause, G., Kutter, C., Miersch, O., Wasternack, C. 2003. Enzymes of jasmonate biosynthesis occur in tomato sieve elements. *Plant and Cell Physiology* 44: 643–648.

Hause, B., Stenzel, I., Miersch, O., Maucher, H., Kramell, R., Ziegler, J., and Wasternack, C. 2000. Tissue-specific oxylipin signature of tomato flowers: Allene oxide cyclase is highly expressed in distinct flower organs and vascular bundles. *Plant J.* 24: 113–126.

Heath, M.C. 2000. Hypersensitive response-related death. *Plant Molecular Biology* 44: 321–334.

Heitz, T., Widemann, E., Lugan, R., Miesch, L., Ullmann, P., Désaubry, L, et al. 2012. Cytochromes P450 CYP94C1 and CYP94B3 catalyze two successive oxidation steps of the plant hormone jasmonoyl-isoleucine for catabolic turnover. *J. Biol. Chem.*

Hind, S., Malinowski, R., Yalamanchili, R., Stratmann, J.W. 2010. Tissue-type specific systemin perception and the elusive systemin receptor. *Plant Signaling & Behavior* 5: 42–44.

Hind, S.R., Pulliam, S.E., Veronese, P., et al. 2011. The COP9 signalosome controls jasmonic acid synthesis and plant responses to herbivory and pathogens. *The Plant Journal* 65: 480–491.

Hisamatsu, Y., 2005. Oxylipin arabidopsides C and D from *Arabidopsis thaliana*. *Journal of Natural Products*. 68:600.

Hisamatsu, Y., Goto, N., Hasegawa, K., and Shigemori, H., 2003. Arabidopsides A and B, two new oxylipins from *Arabidopsis thaliana*. *Tetrahedron letters* 44:5553-5556.

Hofmann, E., Pollmann, S., 2008. Molecular mechanism of enzymatic allene oxide cyclization in plants. *Plant Physiol. Biochem.* 46: 302–308.

Hofmann, E., Zerbe, P., Schaller, F., 2006. The crystal structure of *Arabidopsis thaliana* allene oxide cyclase: insights into the oxylipin cyclization reaction. *Plant Cell.* 18: 3201–3217.

Howe, G.A., and Jander, G. 2008. Plant immunity to insect herbivores. *Annu Rev. Plant Biol.* 59:41-66.

Howe, G.A., Lee, G.I., Itoh, A., Li, L., and DeRocher, A.E. 2000. Cytochrome P450-dependent metabolism of oxylipins in tomato. Cloning and expression of allene oxide synthase and fatty acid hydroperoxide lyase. *Plant Physiol.* 123:711-724.

Howe, G.A., Schilmiller, A.L. 2002. Oxylipin metabolism in response to stress. *Curr. Opin. Plant Biol.* 5:230–236.

Hughes, R.K., Lawson, D.M., Hornostaj, A.R., Fairhurst, S.A., and Casey, R., 2001. Mutagenesis and modelling of linoleate-binding to pea seed lipoxygenase. *Eur. J. Biochem.* 268:1030-1040.

Hyun, Y., Choi, S., Hwang, H.J., Yu, J., Nam, S.J., Ko, J. et al. 2008. Cooperation and functional diversification of two closely related galactolipase genes for jasmonate biosynthesis. *Dev. Cell.* 14:183-192.

Ishiguro, S., Kawai-Oda, A., Ueda, J., Nishida, I., and Okada, K. 2001. The *DEFECTIVE IN ANTHER DEHISCENCE1* gene encodes a novel phospholipase A1 catalyzing the initial step of jasmonic acid biosynthesis, which synchronizes pollen maturation, anther dehiscence, and flower opening in Arabidopsis. *Plant Cell* 13:2191-2209.

Jacinto, T., McGurl, B., Franceschi, V., Delano-Freier, J., Ryan, C.A., 1997. Tomato prosystemin promoter confers wound-inducible, vascular bundle-specific expression of the b-glucuronidase gene in transgenic tomato plants. *Planta* 203: 406-412.

Jayaraj, J., Muthukrishnan, S., Liang, G.H., Velazhahan, R. 2004. Jasmonic acid and salicylic acid induce accumulation of β-1,3-glucanase and thaumatin-like proteins in wheat and enhance resistance against *Stagonospora nodorum*. *Biol. Plant.* 48(3): 425–430.

Jung, C., Yeu, S.Y., Koo, Y.J., Kim, M., Do Choi, Y., Cheong, J.J., 2007. Transcript profile of transgenic Arabidopsis constitutively producing methyl jasmonate. *J. Plant Biol.* 50: 12–17.

Kachroo, P., Yoshioka, K., Shah, J., Dooner, K.D., Klessig, D.F. 2000. Resistance to turnip crinkle virus in Arabidopsis is regulated by two host genes and is salicylic acid dependent but NPR1, ethylene, and jasmonate independent. Plant Cell 12**:** 677–690.

Kandoth, P.K., Ranf, S., Pancholi, S.S., et al. 2007. Tomato MAPKs LeMPK1, LeMPK2, and LeMPK3 function in the systemin-mediated defense response against herbivorous insects. *Proceedings of the National Academy of Sciences of the* USA 104: 12205–12210.

Kang, S., Kim, H.B., Lee, H., et al. 2006. Overexpression in Arabidopsis of a plasma membrane-targeting glutamate receptor from small radish increases glutamate-mediated Ca2+ influx and delays fungal infection. *Molecules and Cells* 21: 418–427.

Katsir, L., Schilmiller, A.L., Staswick, P.E., He, S.Y., and Howe, G.A. 2008. COI1 is a critical component of a receptor for jasmonate and the bacterial virulence factor coronatine. *Proc. Natl. Acad. Sci.* USA 105:7100-7105.

Kazan, K., and Manners, J.M. 2012. JAZ repressors and the orchestration of phytohormone crosstalk. *Trends in Plant Science* 17: 22-31.

Kazan, K., Manners, J.M. 2008. Jasmonate signaling: toward an integrated view. *Plant Physiol.* 146:1459–1468.

Kidd, B.N., Edgar, C.I., Kumar, K.K., Aitken, E.A., Schenk, P.M., Manners, J.M., Kazan, K. 2009. The mediator complex subunit PFT1 is a key regulator of jasmonate-dependent defense in *Arabidopsis*. *Plant Cell.* 21: 2237 – 2252.

Kienow, L., Schneider, K., Bartsch, M., Stuible, H.-P., Weng, H., Miersch, O., Wasternack, C., Kombrink, E., 2008. Jasmonates meet fatty acids: functional analysis of a new acyl-coenzyme A synthetase family from *Arabidopsis thaliana*. *J. Exp. Bot.* 59: 403–419.

Kim, C.Y., Liu, Y., Thorne, E.T., Yang, H., Fukushige, H., Gassmann, W., Hildebrand, D., Sharp, R.E., and Zhang, S. 2003. Activation of a stress-responsive mitogen-activated protein kinase cascade induces the biosynthesis of ethylene in plants. Plant Cell 15: 2707–2718.

Kim, E.S., Kim, H., Park, R.D., Lee, Y., and Han, O. 2002. Dual positional specificity of wound responsive lipoxygenase from maize seedlings. *Journal of Plant Physiology* 159:1263- 1265.

Koch, T., Krumm, T., Jung, V., Engelberth, J., Boland, W. 1999. Differential induction of plant volatile biosynthesis in the lima bean by early and late intermediates of the octadecanoid-signaling pathway. *Plant Physiology* 121: 153–162.

Kombrink, E. 2012. Chemical and genetic exploration of jasmonate biosynthesis and signaling paths. *Planta.* 236(5):1351-66.

Koo, A.J., Chung, H.S., Kobayashi, Y., and Howe, G.A., 2006. Identification of a peroxisomal acyl activating enzyme involved in the biosynthesis of jasmonic acid in Arabidopsis. *J. Biol. Chem.* 281:33511-33520.

Koo, A.J., Cooke, T.F., and Howe, G.A. 2011. Cytochrome P450 CYP94B3 mediates catabolism and inactivation of the plant hormone jasmonoyl-L-isoleucine. *Proc. Natl. Acad. Sci. U.S.A.* 108, 9298–9303.

Koo, A.J.K., Gao, X. L., Jones, A. D., and Howe, G. A. 2009. A rapid wound signal activates the systemic synthesis of bioactive jasmonates in *Arabidopsis*. *Plant J.* 59: 974–986.

Koo, A.J.K., Howe, G.A. 2009. The wound hormone jasmonate. *Phytochemistry* 70: 1571–1580.

Kourtchenko, O., Anderson, M.X., Hamberg, M., Brunnstrom, A., Gobel, C., McPhail, K.L., Gerwick, W.H., Feussner, I., and Ellerstrom, M. 2007. Oxo-phytodienoic acid-containing galactolipids in Arabidopsis: jasmonate signaling dependence. *Plant Physiol.* 145:1658- 1669.

Kudla, J., Batistic, O., Hashimoto, K., 2010. Calcium signals: the lead currency of plant information processing. *The Plant Cell* 22: 541–563.

Kunkel, B.N., and Brooks, D.M. 2002. Cross talk between signaling pathways in pathogen defense. *Curr Opin Plant Biol.* 5: 325-331.

Laudert, D., and Weiler, E.W. 1998. Allene oxide synthase: a major control point in *Arabidopsis thaliana* octadecanoid signalling. *Plant J.* 15:675-684.

Laudert, D., Hennig, P., Stelmach, B.A., Müller, A., Andert, L., Weiler, E.W., 1997. Analysis of 12-oxo-phytodienoic acid enantiomers in biological samples by capillary gas chromatography-mass spectrometry using cyclodextrin stationary phases. *Anal. Biochem.* 246: 211–217.

Laudert, D., Pfannschmidt, U., Lottspeich, F., Hollander-Czytko, H., and Weiler, E.W. 1996. Cloning, molecular and functional characterization of Arabidopsis thaliana allene oxide synthase (CYP 74), the first enzyme of the octadecanoid pathway to jasmonates. *Plant Mol Biol.* 31:323-335.

Laudert, D., Schaller, F., and Weiler, E.W. 2000. Transgenic *Nicotiana tabacum* and *Arabidopsis thaliana* plants overexpressing allene oxide synthase. *Planta* 211:163-165.

Leon, J., and Sanchez-Serrano, J.J. 1999. Molecular biology of jasmonic acid biosynthesis in plants. *Plant Physiol and Biochem.* 37: 373-380.

Levin, E.J., Kondrashov, D.A., Wesenberg, G.E., Phillips, G.N., 2007. Ensemble refinement of protein crystal structures: validation and application. Structure 15: 1040–1052.

Levy, M., Wang, Q., Kaspi, R., Parrella, M.P., Abel, S. 2005. Arabidopsis IQD1, a novel calmodulin-binding nuclear protein, stimulates glucosinolate accumulation and plant defense. *Plant J.* 43: 79–96.

Li, C., Liu, G., Xu, C., Lee, G.I., Bauer, P., Ling, H.Q., Ganal, M.W., Howe, G.A. 2003. The tomato suppressor of prosystemin-mediated responses gene encodes a fatty acid desaturase required for the biosynthesis of jasmonic acid and the production of a systemic wound signal for defense gene expression. *Plant Cell* 15: 1646-1661.

Li, C., Schilmiller, A.L., Liu, G., Lee, G.I., Jayanty, S., Sageman, C., Vrebalov, J., Giovannoni, J.J., Yagi, K., Kobayashi, Y., and Howe, G.A. 2005. Role of ß-oxidation in jasmonate biosynthesis and systemic wound signaling in tomato. *Plant Cell* 17: 971.

Li, L., Li, C., Lee, G., Howe, G. 2002. Distinct roles for jasmonate synthesis and action in the systemic wound response of tomato. *Proceedings of the National Academy of Sciences* of the USA 99: 6416–6421.

Li, L., Zhao, Y., McCaig, B.C., Wingerd, B.A., Wang, J., Whalon, M.E., Pichersky, E., Howe, G.A. 2004. The tomato homolog of CORONATINE-INSENSITIVE1 is required for the maternal control of seed maturation, jasmonate-signaled defense responses, and glandular trichome development. *Plant Cell.* 16:126-143.

Liu, Y.L., Ahn, J.E., Datta, S., Salzman, R.A., Moon, J., Huyghues-Despointes, B., Pittendrigh, B., Murdock, L.L., Koiwa, H., and Zhu-Salzman, K. 2005. Arabidopsis vegetative storage protein is an anti-insect acid phosphatase. *Plant Physiol.* 139:1545-1556.

Loake, G., Grant, M. 2007. Salicylic acid in plant defence - the players and protagonists. *Current Opinion in Plant Biology* 10: 466-472.

Lorenzo, O., and Solano, R. 2005. Molecular players regulating the jasmonate signalling network. *Curr Opin Plant Biol.* 8:532-540.

Lorenzo, O., Chico, J.M., Sanchez-Serrano, J.J., and Solano, R. 2004. JASMONATE-INSENSITIVE1 encodes a MYC transcription factor essential to discriminate between different jasmonate-regulated defense responses in Arabidopsis. *Plant Cell.* 16:1938-1950.

Lorenzo, O., Piqueras, R., Sanchez-Serrano, J.J., and Solano, R. 2003. ETHYLENE RESPONSE FACTOR1 integrates signals from ethylene and jasmonate pathways in plant defense. *Plant Cell* 15:165-178.

Lozano-Duran, R., Rosas-Diaz, T., Gusmaroli, G., Luna, A.P., Taconnat, L., Deng, X.W., and Bejarano, E.R. 2011. Gemini viruses subvert ubiquitination by altering CSN-mediated de-rubylation of SCF E3 ligase complexes and inhibit jasmonate signaling in *Arabidopsis thaliana*. *Plant Cell* 23:1014-1032.

Mason, H.S., Dewald, D.B., Creelman, R.A., Mullet, J.E. 1992. Coregulation of soybean vegetative storage protein gene expression by methyl jasmonate and soluble sugars. Plant Physiol. 98 (3):859–867.

Masterson, C., and Wood, C. 2001. Mitochondrial and peroxisomal beta-oxidation capacities of organs from a non-oilseed plant. *Proc. Biol. Sci.* 268:1949-1953.

Matsui, H., Nakamura, G., Ishiga, Y., Toshima, H., Inagaki, Y., Toyoda, K., Shiraishi, T., and Ichinose, Y. 2004. Structure and expression of 12-oxophytodienoate reductase (subgroup I) genes in pea, and characterization of the oxidoreductase activities of their recombinant products. *Molecular Genetics and Genomics* 271:1-10.

Maucher, H., Hause, B., Feussner, I., Ziegler, J., and Wasternack, C. 2000. Allene oxide synthases of barley (*Hordeum vulgare* cv. *Salome*): tissue specific regulation in seedling development. *Plant J.* 21:199-213.

McConn, M., Creelman, R.A., Bell, E., Mullet, J.E., Browse, J. 1997. Jasmonate is essential for insect defense in Arabidopsis. *Proc. Natl. Acad. Sci.* USA 94:5473-5477.

McGrath, K.C., Dombrecht, B., Manners, J.M., Schenk, P.M., Edgar, C.I., Maclean, D.J., Scheible, W.R., Udvardi, M.K., and Kazan, K. 2005. Repressor- and activator-type ethylene response factors functioning in jasmonate signaling and disease resistance identified via a genome-wide screen of Arabidopsis transcription factor gene expression. *Plant Physiol.* 139:949-959.

Melotto, M., Mecey, C., Niu, Y. et al., 2008. A critical role of two positively charged amino acids in the Jas motif of Arabidopsis JAZ proteins in mediating coronatine- and jasmonoyl isoleucine-dependent interactions with the COI1 F-box protein. *Plant J.* 55, 979–988.

Meyer, A., Miersch, O., Buttner, C., Dathe, W., Sembdner, G. 1984. Occurrence of the plant growth regulator jasmonic acid in plants. *J. Plant Growth Regul.* 3:1–8.

Mielke, K., Forner, S., Kramell, R., Conrad, U., Hause, B. 2011. Cell-specific visualization of jasmonates in wounded tomato and Arabidopsis leaves using jasmonate-specific antibodies. *New Phytologist* 190: 1069–1080.

Miersch, O., Porzel, A., and Wasternack, C. 1999. Microbial conversion of jasmonates hydroxylations by *Aspergillus niger*. *Phytochemistry* 50:1147-1152.

Moalem-Beno, D., Tamari, G., Leitner-Dagan, Y., Borochov, A., and Weiss, D. 1997. Sugardependent gibberellin-induced chalcone synthase gene expression in *Petunia corollas*. *Plant Physiol.* 113:419-424.

Moffat, C.S., Ingle, R.A., Wathugala, D.L., Saunders, N.J., Knight H., Knight M.R. 2012. ERF5 and ERF6 play redundant roles as positive regulators of JA/Et-mediated defense against *Botrytis cinerea* in *Arabidopsis*. PLOS ONE 7: e35995.

Mosblech, A, Thurow, C., Gatz, C., Feussner, I., Heilmann, I. 2011. Jasmonic acid perception by COI1 involves inositol polyphosphates in *Arabidopsis thaliana*. *The Plant Journal* 65: 949–957.

Mosblech, A., Feussner, I., Heilmann, I., 2009. Oxylipins: structurally diverse metabolites from fatty acid oxidation. *Plant Physiol. Biochem.* 47, 511–517.

Mueller, M.J. 1997. Enzymes involved in jasmonic acid biosynthesis. *Physiologia Plantarum* 100:653-663.

Mueller, M.J. 2004. Archetype signals in plants: the phytoprostanes. *Curr. Opin. Plant. Biol.* 7: 441–448.

Mueller-Uri, F., Parthier, B., Nover, L. 1988. Jasmonate-induced alteration of gene expression in barley leaf segments analyzed by in-vivo and in-vitro protein synthesis. *Planta* 176:241–247

Mur, L.A.J., Kenton, P., Atzorn, R., Miersch, O., Wasternack, C. 2006. The outcomes of concentration-specific interactions between salicylate and jasmonate signaling include synergy, antagonism, and oxidative stress leading to cell death. *Plant Physiology* 140: 249–262.

Nahar, K., Kyndt, T., De Vleesschauwer, D., Höfte, M., Gheysen, G. 2011. The jasmonate pathway is a key player in systemically induced defense against root knot nematodes in rice. *Plant Physiol.* 157(1):305-316.

Nakano, T., Suzuki, K., Fujimura, T., and Shinshi, H. 2006. Genome-wide analysis of the ERF gene family in Arabidopsis and rice. *Plant Physiol.* 140:411-432.

Nemchenko, A., Kunze, S., Feussner, I., and Kolomiets, M. 2006. Duplicate maize 13- lipoxygenase genes are differentially regulated by circadian rhythm, cold stress, wounding, pathogen infection, and hormonal treatments. *J. Exp. Bot.* 57: 3767–3779.

Neumann, P., Brodhun, F., Sauer, K., et al. 2012. Crystal structures of *Physcomitrella patens* AOC1 and AOC2: insights into the enzyme mechanism and differences in substrate specificity. *Plant Physiology* 160:1251-1266.

Nilsson, A.K., Fahlberg, P., Ellerstro¨m, M., Anderson, M.X. 2012. Oxo-phytodienoic acid (OPDA) is formed on fatty acids esterified to galactolipids after tissue disruption in Arabidopsis thaliana. *FEBS Letters* 586: 2483–2487.

Ohya, H., Ogata, A., Nakamura, K., Chung, K.M., and Sano, H. 2008. A stress-responsive multifunctional protein involved in beta-oxidation in tobacco plants. *Plant Biotechnology* 25:503-508.

Osbourn, A.E. 1996. Preformed antimicrobial compounds and plant defense against fungal attack. *Plant Cell* 8: 1821–1831

Pan, Z.Q., Durst, F., Werckreichhart, D., Gardner, H.W., Camara, B., Cornish, K., and Backhaus, R.A. 1995. The major protein of guayule rubber particles is a cytochrome-p450 - characterization based on cDNA cloning and spectroscopic analysis of the solubilized enzyme and its reaction-products. *Journal of Biological Chemistry* 270:8487-8494.

Paré, P.W., and Tumlinson, J.H. 1999. Plant volatiles as a defense against insect herbivores. *Plant physiology* 121: 325-331.

Pauwels, L., and Goossens, A. 2011. The JAZ proteins: a crucial interface in the jasmonate signaling cascade. *Plant Cell* 23:3089-3100.

Pauwels, L., Morreel, K., De Witte, E., Lammertyn, F., Van Montagu, M., Boerjan, W., Inze, D., and Goossens, A. 2008. Mapping methyl jasmonate-mediated transcriptional reprogramming of metabolism and cell cycle progression in cultured *Arabidopsis* cells. *Proc. Natl. Acad. Sci. U.S.A.* 105: 1380–1385.

Pearce, G., Strydom, D., Johnson, S., Ryan, C. 1991. A polypeptide from tomato leaves induces wound-inducible proteinase inhibitor proteins. *Science* 253: 895–897.

Peever, T.L. and Higgins, V.J. 1989. Electrolyte leakage, lipoxygenase, and lipid peroxidation induced in tomato leaf tissue by specific and nonspecific elicitors from *Cladosporium fulvum*. *Plant physiology* 90:867-875.

Piel, J., Atzorn, R., Gäbler, R., Kühnemann, F., Boland, W. 1997. Cellulysin from the plant parasitic fungus *Trichoderma viride* elicits volatile biosynthesis in higher plants via the octadecanoid signaling cascade. *FEBS Letters* 416: 143–148.

Pieterse, C.M., Van Wees, S.C., Van Pelt, J.A., Knoester, M., Laan, R., Gerrits, H., Weisbeek, P.J., and Van Loon, L.C. 1998. A novel signaling pathway controlling induced systemic resistance in Arabidopsis. Plant Cell 10:1571-1580.

Porta, H., and Rocha-Sosa, M. 2002. Plant lipoxygenases. Physiological and molecular features. *Plant Physiol.* 130:15-21.

Pré, M., Atallah, M., Champion, A., De Vos, M., Pieterse, C.M., and Memelink, J. 2008. The AP2/ERF domain transcription factor ORA59 integrates jasmonic acid and ethylene signals in plant defense. *Plant Physiol.* 147:1347-1357.

Quina, F.H., Moreira, P.F., Vautier-Giongo, C., Rettori, D., Rodrigues, R.F., Freitas, A.A., Silva, P.F., and Macanita, A.L.2009. Photochemistry of anthocyanins and their biological role in plant tissues. *Pure and Applied Chemistry* 81:1687-1694.

Raguso, R.A. and Pichersky, E. 1995. Floral volatiles from *Clarkia breweri* and *C. concinna* (Onagraceae): recent evolution of floral scent and moth pollination. *Plant Syst. Evolu.* 194:55-67.

Ralph, S., Oddy, C., Cooper, D., Yueh, H., Jancsik, S., Kolosova, N., et al. 2006. Genomics of hybrid poplar (*Populus trichocarpa* x *deltoides*) interacting with forest tent caterpillars (*Malacosoma disstria*): normalized and full-length cDNA libraries, expressed sequence tags, and a cDNA microarray for the study of insect-induced defenses in poplar. *Mol. Ecol.* 15: 1275–1297.

Reymond, P., Bodenhausen, N., Van Poecke, R.M., Krishnamurthy, V., Dicke, M., and Farmer, E.E. 2004. A conserved transcript pattern in response to a specialist and a generalist herbivore. *Plant Cell* 16: 3132-3147.

Reymond, P., Weber, H., Damond, M., and Farmer, E.E. 2000. Differential gene expression in response to mechanical wounding and insect feeding in *Arabidopsis*. *Plant Cell* 12: 707–720.

Robert, A. Raguso, Douglas M. Light, Eran Pickersky 1996. Electroantennogram responses of *Hyles lineata* (Sphingidae: Lepidoptera) to volatile compounds from *Clarkia breweri* (Onagraceae) and other moth-pollinated flowers. Journal of Chemical Ecology October 1996, Volume 22, Issue 10, pp 1735-1766.

Ryals, J., Ward, E., and Métraux, J.P. 1992. Systemic acquired resistance: An inducible defense mechanism in plants. In lnducible Plant Proteins: Their Biochemistry and Molecular Biology, J.L. Wray, Ed (Cambridge, UK: Cambridge University Press), pp. 205-229.

Ryan, C.A. 2000. The systemin signaling pathway: differential activation of plant defensive genes. *Biochimica et Biophysica Acta* 1477: 112–121.

Sanders, P.M., Lee, P.Y., Biesgen, C., Boone, J.D., Beals, T.P., Weiler, E.W., Goldberg, R.B., 2000. The Arabidopsis DELAYED DEHISCENCE1 gene encodes an enzyme in the jasmonic acid synthesis pathway. *Plant Cell* 12: 1042–1061.

Saniewski, M., J. Czapski, 1983. The effect of methyl jasmonate on lycopene and b-carotene accumulation in ripening red tomato. *Experientia*. 39:1373-1374

Sasaki, Y., Asamizu, E., Shibata, D., Nakamura, Y., Kaneko, T., Awai, K., Amagai, M., et al., 2001. Monitoring of methyl jasmonate-responsive genes in Arabidopsis by cDNA macroarray: self-activation of jasmonic acid biosynthesis and crosstalk with other phytohormone signaling pathways. *DNA Res.* 8:153-161

Sasaki-Sekimoto, Y., Taki, N., Obayashi, T., Aono, M., Matsumoto, F., Sakurai, N., Suzuki, H., et al., 2005. Coordinated activation of metabolic pathways for antioxidants and defence compounds by jasmonates and their roles in stress tolerance in *Arabidopsis*. *Plant J.* .44: 653–668.

Schaller, A., and Stintzi, A. 2009. Enzymes in jasmonate biosynthesis - structure, function, regulation. *Phytochemistry* 70:1532-1538.

Schaller, F., 2001. Enzymes of the biosynthesis of octadecanoid-derived signalling molecules. *J. Exp. Bot.* 52:11-23.

Schaller, F., Biesgen, C., Müssig, C., Altmann, T., Weiler, E.W., 2000. 12-Oxophytodienoate reductase 3 (OPR3) is the isoenzyme involved in jasmonate biosynthesis. *Planta* 210: 979–984.

Schaller, F., Hennig, P., and Weiler, E.W. 1998. 12-Oxophytodienoate-10,11-reductase: occurrence of two isoenzymes of different specificity against stereoisomers of 12- oxophytodienoic acid. *Plant Physiol.* 118:1345-1351.

Schaller, F., Schaller, A., Stintzi, A. 2004. Biosynthesis and metabolism of jasmonates. *J. Plant Growth Regul.* 23:179–199.

Schaller, F., Weiler, E.W. 1997. Molecular cloning and characterization of 12-oxophytodienoate reductase, an enzyme of the octadecanoid signaling pathway from *Arabidopsis thaliana*. Structural and functional relationship to yeast old yellow enzyme. *J. Biol. Chem.* 272:28066–28072.

Schenk, P.M., Kazan, K., Wilson, I., Anderson, J.P., Richmond, T., Somerville, S.C., Manners, J.M. 2000. Coordinated plant defense responses in Arabidopsis revealed by microarray analysis. *Proc. Natl. Acad. Sci. USA* 97**:** 11655–11660.

Schilmiller, A.L., Howe, G.A. 2005. Systemic signaling in the wound response. *Curr. Opin. Plant Biol.* 8:369–377.

Schilmiller, A.L., Koo, A.J., and Howe, G.A. 2007. Functional diversification of acyl-coenzyme A oxidases in jasmonic acid biosynthesis and action. *Plant Physiol.* 143:812-824.

Schneider, K., Kienow, L., Schmelzer, E., Colby, T., Bartsch, M., Miersch, O., Wasternack, C., Kombrink, E., Stuible, H.P., 2005. A new type of peroxisomal acyl-coenzyme A synthetase from *Arabidopsis thaliana* has the catalytic capacity to activate biosynthetic precursors of jasmonic acid. *J. Biol. Chem.* 280: 13962– 13972.

Schommer, C., Palatnik, J.F., Aggarwal, P., Chetelat, A., Cubas, P., Farmer, E.E., Nath, U., and Weigel, D. 2008. Control of jasmonate biosynthesis and senescence by miR319 targets. *PLoS Biol.* 6:e230.

Sembdner, G., and Parthier, B. 1993. The Biochemistry and the Physiological and Molecular Actions of Jasmonates. *Annu. Rev. Plant. Physiol. and Plant Mol Biol.* 44:569-589.

Seo, H.S., Song, J.T., Cheong, J.J., Lee, Y.H., Lee, Y.W., Hwang, I., Lee, J.S., and Choi, Y.D., 2001. Jasmonic acid carboxyl methyltransferase: a key enzyme for jasmonate-regulated plant responses. *Proc. Natl. Acad. Sci.* U S A 98:4788-4793.

Seo, S., Okamoto, M., Seto, H., Ishizuka, K., Sano, H. and Ohashi, Y. 1995. Tobacco MAP kinase: a possible mediator in wound signal transduction pathways. *Science*, 270: 1988–1992.

Shan, X., Zhang, Y., Peng, W., Wang, Z., and Xie, D. 2009. Molecular mechanism for jasmonate induction of anthocyanin accumulation in Arabidopsis. *J. Exp. Bot.* 60: 3849-3860.

Shang, J., Xi, D.H., Xu, F., Wang, S.D., Cao, S., Xu, M.Y., Zhao, P.P., Wang, J.H., Jia, S.D., Zhang, Z.W., Yuan, S., and Lin, H.H. 2011. A broad-spectrum, efficient and non-transgenic approach to control plant viruses by application of salicylic acid and jasmonic acid. *Planta* 233: 299- 308.

Sheard, L.B., Tan, X., Mao, H., Withers, J., Ben-Nissan, G., Hinds, T.R., Kobayashi, Y., et al., 2010. Jasmonate perception by inositol-phosphate-potentiated COI1-JAZ co-receptor. *Nature* 468:400-405.

Shin, J.H., Van, K., Kim, D.H., Kim, K.D., Jang, Y.E., Choi, B.S., Kim, M.Y., and Lee, H. 2008. The lipoxygenase gene family: a genomic fossil of shared polyploidy between *Glycine max* and *Medicago truncatula. BMC Plant Biol.* 8:133.

Shockey, J.M., Fulda, M.S., Browse, J., 2003. Arabidopsis contains a large superfamily of acyl-activating enzymes. Phylogenetic and biochemical analysis reveals a new class of acyl-coenzyme a synthetases. *Plant Physiol.* 132: 1065–1076.

Shoji, T., and Hashimoto, T. 2011. Tobacco MYC2 regulates jasmonate-inducible nicotine biosynthesis genes directly and by way of the NIC2-locus ERF genes. *Plant Cell Physiol.* 52: 1117-1130.

Shoji, T., Ogawa, T., and Hashimoto, T. 2008. Jasmonate-induced nicotine formation in tobacco is mediated by tobacco COI1 and JAZ genes. *Plant Cell Physiol.* 49:1003-1012.

Shyu, C., Figueroa, P., DePew, C.L. et al. 2012. JAZ8 lacks a canonical degron and has an EAR motif that mediates transcriptional repression of jasmonate responses in Arabidopsis. *The Plant Cell* 24: 536–550.

Smith, J.L., De Moraes, C.M., and Mescher, M.C. 2009. Jasmonate- and salicylate-mediated plant defense responses to insect herbivores, pathogens and parasitic plants. *Pest. Manag Sci.* 65:497-503.

Song, W.C., Funk, C.D., and Brash, A.R. 1993. Molecular cloning of an allene oxide synthase: a cytochrome P450 specialized for the metabolism of fatty acid hydroperoxides. *Proc. Natl. Acad. Sci.* USA 90:8519-8523.

Staswick, P.E., and Tiryaki, I., 2004. The oxylipin signal jasmonic acid is activated by an enzyme that conjugates it to isoleucine in Arabidopsis. *Plant Cell* 16:2117-2127.

Staswick, P.E., Su, W., and Howell, S.H. 1992. Methyl jasmonate inhibition of root growth and induction of a leaf protein are decreased in an Arabidopsis thaliana mutant. *Proc. Natl. Acad. Sci.* USA 89:6837-6840.

Staswick, P.E., Tiryaki, I., and Rowe, M.L., 2002. Jasmonate response locus *JAR1* and several related Arabidopsis genes encode enzymes of the firefly luciferase superfamily that show activity on jasmonic, salicylic, and indole-3-acetic acids in an assay for adenylation. *Plant Cell* 14:1405-1415.

Staswick, P.E., Yuen, G.Y., and Lehman, C.C. 1998. Jasmonate signaling mutants of Arabidopsis are susceptible to the soil fungus *Pythium irregulare*. *Plant J.* 15: 747-754.

Stelmach, B.A., Muller, A., Hennig, P., Gebhardt, S, Schubert-Zsilavecz, M., and Weiler, E.W., 2001. A novel class of oxylipins, sn1-O-(12-oxophytodienoyl)-sn2-O-(hexadecatrienoyl)- monogalactosyl Diglyceride, from *Arabidopsis thaliana*. *J. Biol. Chem.* 276:12832-12838.

Stenzel, I., Hause, B., Maucher, H., et al. 2003a. Allene oxide cyclase dependence of the wound response and vascular bundle-specific generation of jasmonates in tomato – amplification in wound signaling. *The Plant Journal* 33: 577–589.

Stenzel, I., Hause, B., Miersch, O., Kurz, T., Maucher, H., Weichert, H., Ziegler, J., Feussner, I., and Wasternack, C., 2003. Jasmonate biosynthesis and the allene oxide cyclase family of *Arabidopsis thaliana*. *Plant Mol Biol.* 51:895-911.

Stenzel, I., Otto, M., Delker, C., et al. 2012. ALLENE OXIDE CYCLASE (AOC) gene family members of *Arabidopsis thaliana*: tissue- and organ-specific promoter activities and in vivo heteromerization. *Journal of Experimental Botany* 63: 6125– 6138.

Stintzi, A., Browse, J., 2000. The Arabidopsis male-sterile mutant, opr3, lacks the 12- oxophytodienoic acid reductase required for jasmonate synthesis. *Proc. Natl. Acad. Sci.* USA 97: 10625–10630.

Stintzi, A., Weber, H., Reymond, P., Browse, J., and Farmer, E.E. 2001. Plant defense in the absence of jasmonic acid: the role of cyclopentenones. *Proc. Natl. Acad Sci.* USA 98: 12837-12842.

Strassner, J., Schaller, F., Frick, U.B., Howe, G.A., Weiler, E.W., Amrhein, N.A., Macheroux, P., Schaller, A., 2002. Characterization and cDNA-microarray expression analysis of 12-oxophytodienoate reductases reveals differential roles for octadecanoid biosynthesis in the local versus the systemic wound response. *Plant J.* 32: 585–601.

Stumpe, M., and Feussner, I. 2006. Formation of oxylipins by CYP74 enzymes. *Phytochem. Rev.* 5:347-357.

Sun, J.Q., Jiang, H.L., Li, C.Y. 2011a. Systemin/jasmonate-mediated systemic defense signaling in tomato. *Molecular Plant* 4: 607–615.

Suza, W.P., and Staswick, P.E. 2008. The role of JAR1 in Jasmonoyl-L: -isoleucine production during Arabidopsis wound response. *Planta* 227:1221-1232.

Suza, W.P., Rowe, M.L., Hamberg, M., and Staswick, P.E. 2010. A tomato enzyme synthesizes (+)-7-iso-jasmonoyl-L-isoleucine in wounded leaves. *Planta* 231: 717–728.

Takahashi, F., Yoshida R, Ichimura K, Mizoguchi T, Seo S, Yonezawa M, et al. 2007. The mitogen-activated protein kinase cascade MKK3-MPK6 is an important part of the jasmonate signal transduction pathway in Arabidopsis. *The Plant Cell* 19: 805-818.

Taki, N., Sasaki-Sekimoto, Y., Obayashi, T., Kikuta, A., Kobayashi, K., Ainai, T., Yagi, K., Sakurai, N., et al., 2005. 12-Oxo-phytodienoic acid triggers expression of a distinct set of genes and plays a role in wound-induced gene expression in Arabidopsis. *Plant Physiol.* 139: 1268–1283.

Tamogami, S., Ralkwal, R., and Agrawal, G.K. 2008. Interplant communication: airborne methyl jasmonate is essentially converted into JA and JA-Ile activating jasmonate signaling pathway and VOCs emission. *Biochem. Biophys. Res. Commun.* 376: 723–727.

Tani, T., Sobajima, H., Okada, K., Chujo, T., Arimura, S., Tsutsumi, N., Nishimura, M., Seto, H., Nojiri, H., and Yamane, H. 2008. Identification of the *OsOPR7* gene encoding 12- oxophytodienoate reductase involved in the biosynthesis of jasmonic acid in rice. *Planta* 227:517-526..

Thatcher, L.F., Manners, J.M., and Kazan, K. 2009. *Fusarium oxysporum* hijacks COI1-mediated jasmonate signaling to promote disease development in Arabidopsis. *Plant J.* 58:927-939.

Theodoulou, F.L., Job, K., Slocombe, S.P., Footitt, S., Holdsworth, M., Baker, A., Larson, T.R., Graham, I.A., 2005. Jasmonic acid levels are reduced in COMATOSE ATP binding cassette transporter mutants. Implications for transport of jasmonate precursors into peroxisomes. *Plant Physiol.* 137: 835–840.

Thines, B., Katsir, L., Melotto, M., Niu, Y., Mandaokar, A., Liu, G., Nomura, K., He, S.Y., Howe, G.A., and Browse, J., 2007. JAZ repressor proteins are targets of the SCF (COI1) complex during jasmonate signalling. *Nature* 448:661-665.

Thomma, B.P., Eggermont, K., Penninckx, I.A., Mauch-Mani, B., Vogelsang, R., Cammue, B.P., and Broekaert, W.F. 1998. Separate jasmonate-dependent and salicylate-dependent defense response pathways in Arabidopsis are essential for resistance to distinct microbial pathogens. *Proc. Natl. Acad. Sci. USA* 95:15107-15111

Tilton, G.B., Shockey, J.M., and Browse, J. 2004. Biochemical and molecular characterization of ACH2, an acyl-CoA thioesterase from *Arabidopsis thaliana*. *J. Biol. Chem.* 279:7487-7494.

Ueda, J., and Kato, J. 1980. Isolation and Identification of a Senescence-promoting Substance from Wormwood (*Artemisia absinthium* L.). *Plant Physiol.* 66:246-249.

Umate, P. 2011. Genome-wide analysis of lipoxygenase gene family in Arabidopsis and rice. *Plant Signal Behav*. 6:335-338.

Utsunomiya, Y., Nakayama, T., Oohira, H., Hirota, R., Mori, T., Kawai, F., and Ueda, T. 2000. Purification and inactivation by substrate of an allene oxide synthase (CYP74) from corn (*Zea mays* L.) seeds. *Phytochemistry* 53:319-323.

Van Loon, L.C., Rep, M., Pieterse, C.M.J. 2006. Significance of inducible defense-related proteins in infected plants. *Annu. Rev. Phytopath* 44: 135–162.

Van Verk, M., Bol, J., Linthorst, H., 2011. Prospecting for genes involved in transcriptional regulation of plant defenses, a bioinformatics approach. *BMC Plant Biology* 11: 88.

Vanholme, B., Grunewald, W., Bateman, A., Kohchi, T., Gheysen, G. 2007. The tify family previously known as ZIM. *Trends in Plant Science* 12: 239– 244.

Vick, B.A., and Zimmerman, D.C. 1983. The biosynthesis of jasmonic acid: a physiological role for plant lipoxygenase. *Biochem Biophys Res Commun.* 111:470-477.

Vijayan, P., Shockey, J., Lévesque, C.A., Cook, R.J., and Browse, J. 1998. A role for jasmonate in pathogen defense of Arabidopsis. *Proc Natl Acad Sci*. USA 95:7209-7214.

Wager, A., Browse, J. 2012. Social Network: JAZ protein interactions expand our knowledge of jasmonate signaling. *Frontiers in Plant Science* 3: 41.

Walter, A., Mazars, C., Maitrejean, M., Hopke, J., Ranjeva, R., Boland, W., Mithöfer, A. 2007. Structural requirements of jasmonates and synthetic analogues as inducers of Ca2+ signals in the nucleus and the cytosol of plant cells. *Angew. Chem. Int. Ed Engl.* 46: 4783–4785

Wasternack, C. 2006. Oxylipins: biosynthesis, signal transduction and action. In: Hedden P, Thomas S eds. Plant Hormone Signaling. *Harpenden*, Blackwell Publishing, 185–228.

Wasternack, C., 2007. Jasmonates: an update on biosynthesis, signal transduction and action in plant stress response, growth and development. *In Ann Bot*, pp 1–17.

Wasternack, C., and B. Hause, 2013. Jasmonates: biosynthesis, perception, signal transduction and action in plant stress response, growth and development. *Annals of Botany* Page 1 of 38.

Wasternack, C., and Hause, B., 2013. Jasmonates: biosynthesis, perception, signal transduction and action in plant stress response, growth and development. *Annals of Botany*, Page 1 of 38.

Wasternack, C., Hause, B. 2002. Jasmonates and octadecanoids – signals in plant stress response and development. In: Moldave K. ed. *Progress in nucleic acid research and molecular biology*. New York, Academic Press, 165–222.

Wasternack, C., Kombrink, E. 2010. Jasmonates: Structural requirements for lipid-derived signals active in plant stress responses and development. *ACS Chemical Biology* 5: 63–77.

Weidhase, R.A., Kramell, H.M., Lehmann, J., Liebisch, H.W., Lerbs, W., Parthier, B. 1987. Methyl jasmonate-induced changes in the polypeptide pattern of senescing barley leaf segments. *Plant Sci.* 51:177–186.

Werck-Reichhart, D., Bak, S., Paquette, S., 2002. Cytochromes P450. In: The Arabidopsis Book. *American Society of Plant Biologists*, Rockville, MD.

Wu, J., Hettenhausen, C., Meldau, S., Baldwin, I.T. 2007. Herbivory rapidly activates MAPK signaling in attacked and unattacked leaf regions but not between leaves of *Nicotiana attenuata*. *The Plant Cell* 19: 1096–1122.

Wu, J.S., Wang, L., and Baldwin, I.T. 2008. Methyl jasmonate-elicited herbivore resistance: does MeJA function as a signal without being hydrolyzed to JA? *Planta*. 227: 1161–1168.

Xie, D.X., Feys, B.F., James, S., Nieto-Rostro, M., Turner, J.G. 1998. *COI1*: an *Arabidopsis* gene required for jasmonate-regulated defense and fertility. *Science* 280:1091–1094.

Xu, L.H., Liu, F.Q., Lechner, E., Genschik, P., Crosby, W.L., Ma, H., Peng, W., Huang, D.F., Xie, D.X. 2002. The SCFCOl1 ubiquitin-ligase complexes are required for jasmonate response in *Arabidopsis*. *Plant Cell* 14:1919–1935.

Xu, S., Cooper, A., Sturgill-Koszycki, S., van Heyningen, T., Chatterjee, D., Orme, I., Allen, P. and Russell, D. G. 1994. Intracellular trafficking in *Mycobacterium tuberculosis* and *Mycobacterium avium*-infected macrophages. *J. Immunol.* 153: 2568-2578.

Xu, Y., Chang, P.F., Liu, D., Narasimhan, M.L., Raghothama, K.G., Hasegawa, P.M., and Bressan, R.A., 1994. Plant defense genes are synergistically induced by ethylene and methyl jasmonate. *Plant Cell* 6:1077-1085.

Yan, Y., Christensen, S., Isakeit, T., Engelberth, J., Meeley, R., Hayward, A., Emery, R.J., and Kolomiets, M.V., 2012. Disruption of *OPR7* and *OPR8* reveals the versatile functions of jasmonic acid in maize development and defense. *Plant Cell* 24:1420-1436.

Yan, Y., Stolz, S., Chételat, A., Reymond, P., Pagni, M., Dubugnon, L., Farmer, E.E. 2007. A downstream mediator in the growth repression limb of the jasmonate pathway. *Plant Cell* 19:2470-2483.

Yang, D.L., Yao, J., Mei, C.S., et al. 2012. Plant hormone jasmonate prioritizes defense over growth by interfering with gibberellin signaling cascade. *Proceedings of the National Academy of Sciences of the USA* 109: E1192–E1200.

Yang, W., Devaiah, S.P., Pan, X., Isaac, G., Welti, R., Wang, X. 2007. AtPLAI is an acyl hydrolase involved in basal jasmonic acid production and *Arabidopsis resistance* to *Botrytis cinerea*. *Journal of Chemical Biology* 282: 18116– 18128.

Zerbe, P., Weiler, E.W., Schaller, F., 2007. Preparative enzymatic solid phase synthesis of cis (+)-12-oxo-phytodienoic acid – physical interaction of AOS and AOC is not necessary. *Phytochemistry* 68: 229–236.

Zhang, J., Simmons, C., Yalpani, N., Crane, V., Wilkinson, H., and Kolomiets, M. 2005. Genomic analysis of the 12-oxo-phytodienoic acid reductase gene family of *Zea mays*. *Plant Mol. Biol.* 59:323-343.

Zhou, N., Tootle, T.L., and Glazebrook, J. 1999. Arabidopsis *PAD3*, a gene required for camalexin biosynthesis, encodes a putative cytochrome P450 monooxygenase. *Plant Cell* 11: 2419-2428.

Ziegler, J., Hamberg, M., Miersch, O., and Parthier, B. 1997. Purification and Characterization of Allene Oxide Cyclase from Dry Corn Seeds. *Plant Physiol.* 114:565-573.

Ziegler, J., Wasternack, C., and Hamberg, M. 1999. On the specificity of allene oxide cyclase. *Lipids* 34:1005-1015.

INDEX

A

B

C

D

E

F

G

H

I

J

K

L

M

N

O

P

Q

R

S

T